MECHANICAL *and* QUARTZ WATCH REPAIR

Mick Watters FBHI

The Crowood Press

First published in 1999 by
The Crowood Press Ltd
Ramsbury, Marlborough
Wiltshire SN8 2HR
enquiries@crowood.com
www.crowood.com

This impression 2021

British Library Cataloguing-in-Publication Data
A catalogue record for this book is available from the British Library.

ISBN 978 1 86126 233 2

Acknowledgements
I am grateful to ETA for kindly allowing me to reproduce the illustrations in Fig. 27 (dial fastener) and Fig. 158; to Seiko UK Ltd for allowing me to reproduce the illustrations seen in Figs. 159, 160, 165, 176, 179 and 180; and to Greiner for their willingness to let me reproduce Figs 54 and 55.

Further, I would like to thank ETA, Seiko UK Ltd, Ronda and Citizen for their generous contribution of resources over a number of years to horological training at St Loye's College, Exeter. I sincerely hope that their investment in training is reflected in customer confidence in their product.

Finally, I would like to thank Peter Mitchell for his help in producing a number of line drawings.

Mick Watters

Typeset by Jean Cussons Typesetting
Diss, Norfolk

Typefaces used: text M Plantin; Labels, Gill Sans

Printed and bound in India by Replika Press Pvt. Ltd.

CONTENTS

INTRODUCTION

This book is written primarily for students taking up mechanical and quartz watch repairing, but also for the many enthusiasts who wish to take a serious and more professional interest in watch repairing. The experienced worker may also benefit by clearing up some of the grey areas such as watch lubrication (particularly with respect to chronographs) and positional time-keeping. Even after forty-five years' experience of watch repairing, thirty of which were teaching, I learned a lot of detail researching this book.

The book was written to stand alone, but at the same time I knew that as watch repairing is a natural progression from clock repairing, many repairers who have *The Clock Repairer's Manual* (by the same author and publisher) would probably wish to have a copy of this book. This led to difficulties with content, but after careful consideration and because of the transferability of knowledge and skills between clock and watch repairing, I decided to minimize on duplication of particular topics to include new material. Please be assured that topics omitted were those not universally practised today.

Much of the text here was written as watches were actually being overhauled, but the principles involved are very largely transferable. The watches chosen were selected because they were seen as representative of that particular type of watch.

1 PRELIMINARIES

Repairing watches is a natural progression from repairing clocks, with much of the knowledge and skills being transferable between the two. It is, however, quite possible to acquire the necessary attributes to repair watches without having first repaired clocks; the work involved is certainly cleaner, and takes up less space. Because it is much smaller work I would encourage the novice to progress slowly: for instance, start by working on larger, simple pocket watches, then take on the smaller size of ladies wrist watches, until eventually you are competent enough to undertake complicated watches of all sizes.

THE WORKING ENVIRONMENT

You will need to work in a clean, quiet, warm, well lit room with minimal distraction, so you can concentrate. If you operate from a shop premises your work bench could be in the front of the shop, though a room removed from, but with easy access to, the front shop is better. The room should be separate from where heavy work is done, and also separate from the area where watch or clock cleaning or manufacturing is carried out.

The Bench

If you are working from home, a bench in a spare room is ideal, then work in progress may be left with little or no effect on other activities in the house – the workroom may even become a happy retreat. The bench could be placed on an existing table, or even better, support it by two sets of drawers of about 28in (71cm) in height; tools and materials could then be kept in these.

Leave enough of a gap between the lowest drawer and the floor to be able to sweep underneath so you can recover the bits and pieces which will inevitably be dropped; it is always the smallest, most crucial parts that get lost, and if the drawers are flat on the floor and the floor surface is even slightly uneven, or the bottom of the drawer cabinet has even the shallowest radius, they are bound to work their way right underneath.

The top of the bench needs to be substantial to prevent bowing, but the other sections could be made of lighter material to reduce the overall costs and weight. The two storage areas under the top are useful anyway, but primarily are necessary to raise the bench to a comfortable working height. In the United Kingdom 36in (91cm) is the usual height for a bench, but in Switzerland 42in (107cm) is quite common, with a chair height that positions the chin only just above the bench top.

In the bench illustrated the material I chose to make the top was MDF (medium-density fibreboard); however, I am bound to point out here the current concerns about working with MDF, as it can cause eye, nose and throat irritations, and more important, it is now thought that it may increase the risk of cancer. Until more is known about the effects of MDF, I would therefore recommend an alternative material such as melamine, or a kitchen work top with a smooth, plain surface. If you do use MDF, take every precaution *not* to breathe in the dust created by working with it, and seal the surface by painting or varnishing.

I decided to avoid the difficulties of anything other than butt-joints, and arranged for a local DIY store to cut the material for me, knowing

Fig. 1 An effective and inexpensive home-made bench.

that they would be able to make flat, square edges with their circular saw or bandsaw, in a way that I would be unlikely to achieve with a hand saw – and good edges would in turn enable glued and screwed butt-joints. It turned out to be a wise move, as the end result was a simple-to-make, rigid bench, where I needed only to drill a few pilot and clearance holes in order to then glue and screw the sections together. After gluing and screwing, the sections remained upright without any additional support.

The following cutting list is for MDF; for other materials it may need to be modified. Also the height of the sides attached to the bench will depend on the height of whatever supports it, be it a table or drawers. A minimum overall height of 36in (91cm) is required, but it may be up to 42in (107cm) for those who prefer it higher. I wanted mine to be 36in, and to achieve this I made the sides 6⅞in (17.5cm) high as it was

going to be placed on a study table 28in (71cm) high.

Top	1	48in × 24in × 18mm
Base	2	20½in × 14in × 12mm
Sides	4	20½in × 6in × 12mm
Back	2	13¼in × 6in × 12mm

Assembly Instructions
1. Begin by wiping over your material with a damp cloth to remove any dust.
2. On a flat surface – perhaps what will be the top of your bench – position two sides, a back and a base to begin assembly.
3. Having made sure the panels are correctly aligned, mark and drill ⅛in (3mm) pilot holes through the base and into the sides to a depth of 1¾in (4.5cm) to take 1¾in (4.5cm) No.8 furniture screws (the type recommended for chipboard furniture panels).

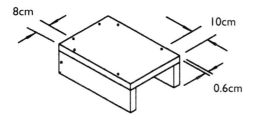

8cm | 10cm | 0.6cm

Fig. 2 Two sides, back and base positioned and marked ready for drilling.

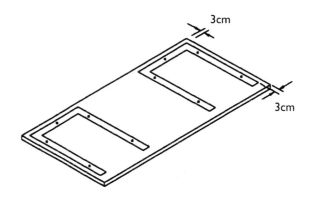

3cm | 3cm

Fig. 3 Underneath side of the bench top marked for drilling.

4. Similarly, drill one central hole in each side panel to secure the back.

5. Remove the sections and enlarge the drilled holes where the screws enter to ¼in (5mm); these are clearance holes.

6. Apply a small quantity of wood glue to the butt-joints, position the panels carefully, and secure with eight screws. The screws are self-countersinking in MDF and should be tightened so they lie just below the surface of the panels.

7. Wipe away any surplus glue with a damp cloth.

8. Repeat steps 2 to 7 for the other support. Allow sufficient time for the glue to harden.

9. Position both supports on the underside of the bench top, and mark where the sides and back make contact. Both back and side panels are 1¹⁄₁₆ in (3cm) in from the edge of the bench top. Identify which is the left and which is the right support.

10. Still working from the back, drill ⅛in (3mm) holes as shown. As they are through-holes, place scrap wood under the bench top: the drill will then hit this, rather than the surface underneath, thereby protecting the latter.

11. Turn both supports over, remembering which is the right and which the left, and align them carefully with the pencil marks. Pass the drill through the ⅛in (3mm) holes already in the bench top, and drill into the supports by a total of 1¾in (4.5cm) (this measurement includes the bench top).

13. Remove the top and increase the size of each of the holes in it to ¼in (5mm) for clearance.

14. Apply a small quantity of wood glue to the top edges of one of the supports, positioning the other simply to hold the bench horizontal.

15. Replace the top, aligning the glued support carefully, and secure with a further six screws each side. Wipe away any surplus glue.

16. Apply glue to the top of the other support, carefully position it, and fix with six screws as before.

The Bench Covering
The top of the bench will need a durable surface covering; it should be plain so that any bits lying on it can be detected easily, and it should be a light, restful colour. I selected a low-cost Formica; I had wanted light green, but an off-cut in light blue was my only choice if I didn't want to incur the cost of a full 8 × 4ft (2.4 × 1.2m) sheet. A high-spec Formica, lino or vinyl would also have been suitable. Having had problems in the past positioning an 'exact fit' bench covering, I ordered one which was a couple of inches over-size all round; it was then an easy task to trim it to a good fit and stick it down with contact adhesive.

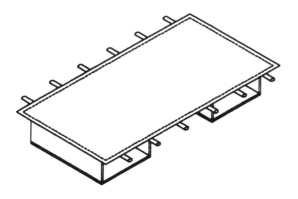

Fig. 4 The bench top, Formica and battens positioned for fixing.

To Fix the Top

1. Have ready half a dozen battens to place on the prepared bench top (actually I used bamboo that I keep for supporting green beans).
2. Apply contact adhesive such as Evo-stik 528 to both surfaces.
3. Position the battens and the Formica.
4. Slide the battens away from the centre without moving the Formica, then press the centre of the Formica down to the bench top.
5. Continue to work from the centre out, moving the battens as necessary until the whole of the Formica is in contact with the bench top.
6. Lay a straight-edge across the Formica in line with the edge of the bench, and score across it with a sharp knife. This is to prevent the Formica from chipping as it is sawn to size.
7. Saw the Formica to size, or up to ¹⁄₁₆in (1.5mm) oversize.

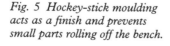

Fig. 5 Hockey-stick moulding acts as a finish and prevents small parts rolling off the bench.

8. File the remaining ¹⁄₁₆in to meet the bench top exactly.
9. To finish the sides and the back of the bench I used a hockey-stick moulding (purchased from any DIY store), to close any gaps and to prevent things rolling off the bench top.
10. The bench was completed by fitting a half-round moulding across the front; this is a more comfortable finish for the wrists which will spend a lot of the time resting on the edge of the bench.

Your Chair

Your work chair should be adjustable, starting from about 16in (41cm) in height, and going up to 24in (61cm) so that you can work comfortably at different heights with a straight back. A dining-room chair can be used, but is less than ideal.

The Bench Light

To light the bench I have tried ordinary tungsten lamps and twin fluorescent lamps, but currently I am using a 60-watt, blue-tinted daylight bulb and find that this gives a very good light to work by. (The credit goes to my wife who used the bulb for cross stitch.)

The Floor

For watch work there is nothing to beat a canvas or vinyl-type floor with a plain surface: inevitably you will drop parts, and with this sort of surface they can be located more easily and swept up. Next to the tweezers and eyeglass, the workshop sweeping brush is one of the most important additions to the workshop.

TOOLS AND CAPITAL EQUIPMENT

Below is a list of recommended tools and capital equipment with which to make a start.

Hand Tools
Tweezers: 1. general purpose
2. fine, for hairspring work

            3. antimagnetic
            4. plastic, for handling batteries
Screwdrivers, set of nine watchmaker's
Material tray
Eyeglasses: 1. double ×10 magnification
            2. single 3in (7.6cm)
               (or equivalent spectacle-type eye-
               glasses if preferred)
Four oilcups
Spirit jar (jam sampler jar)
Brush, soft four-row
Dust blower
Knife: 1. case-opening
       2. craft
Hammer, watchmaker's
Oilers, set of four
Pricker (oiler, tapered to 0.00mm diameter)
Movement holders, set of 12 plastic
Vice stake, 36 hole
Pinchuck, small
Pliers: 1. flat nose
        2. round nose
Top cutters, flush-cutting

In addition to the tools mentioned, you will need a bundle of watch-size pegwood, a small quantity of pith, and some acid-free tissue paper. Any tool and material supplier will help you with any of these.

Capital Equipment

A fundamental requirement is a cleaning machine. These can be bought second-hand, and in fact often come up in horological tool sales; however, for the enthusiast who doesn't want to make a significant commitment but wants a quick start, consider using a jewellery cleaner such as L&R Sonic Jewellery Cleaner. The cost is so low that three might be justified: one for cleaning, one for the first rinse and another for the second rinse. Of course, one could be used for all three processes by simply changing the fluid. Drying could be by ordinary hair dryer. For convenience you would need a suitable wire-mesh basket; this could be selected from a tool catalogue showing spare parts for cleaning machines.

Eventually you will want a staking outfit, a jewelling outfit and a watchmaker's lathe. Again, these can be had second-hand – and be aware that how much you pay for second-hand horological tools varies quite dramatically, so don't feel you have to pay top price. For years I used my staking outfit as a jewelling outfit by making pushers that fitted one particular punch. If you have a choice of staking outfits, get one with an inverto base: each punch can then double as a stake, and it will be much more versatile. And if you buy a second-hand lathe, 8mm is better than 6mm as spare collets will be available. Expect to pay about one third to one half of the new price if tools are in good condition. Old, well used tools can go for significantly less.

Other tools will be identified as they are used, but at least with the above hand tools you will be able to make a modest start at repairing watches.

A FEW PRELIMINARIES

Before we start to overhaul our first watch together it would be useful to understand the following:
• The principles of shock-proofing;
• The main characteristics of watch screws so that the appearance of a screw will tend to point to its likely purpose;
• The characteristics of good screwdrivers;
• What a watch is, and to have a basic knowledge of part names;
• A lubrication chart.

Shock-Proofing the Balance

Let us suppose that a watch is dropped and lands squarely on its back or front. On contact with the ground the balance tends to continue moving, due to its mass, and consequently the end of the pivot strikes the cap jewel with great force; as a result the following may happen: the pivot can mushroom on its end; the cap jewel can smash; or the rounded root of the pivot can smash the jewel hole.

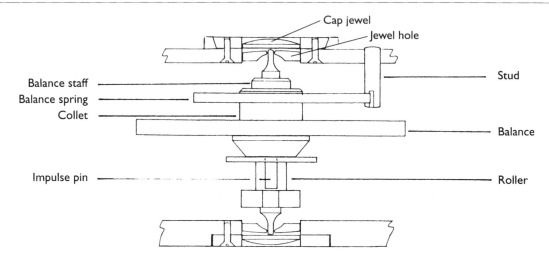

Fig. 6 The complete balance with unprotected jewel hole and cap jewel.

If the watch lands on its edge, the lateral shock causes one or both pivots to break, and may smash one or both jewel holes.

The solution to all these problems is to allow the top and bottom jewel holes and cap jewels the capacity to displace: this they will do momentarily, so that longitudinal shocks are taken by the shoulder behind the pivot, and lateral shocks are absorbed by the diameter behind the pivot. Once the shock is over, the jewel hole and cap jewel return to their original position due to subtle angles on the settings and shock springs. During this displacement, which can happen several times a day, it is quite likely that no time loss is detectable at the hands.

One of the best known systems for shock-proofing is Incabloc made by Portescap.

Handling Incabloc Units
Dismantling

1. With the point of your tweezers, unhook the two free ends of the shock springs (often called lyre springs because of their shape) and hinge them back – though beware, because whilst some are firmly located and cannot 'fly', others are designed to come out easily, and if they do this uncontrolled, it's the last you are likely to see of them. Remove both units, the one in the balance cock and the one in the bottom plate.

2. After removing the jewel holes and cap jewels, re-lock the springs.
3. Separate the jewel holes from the cap jewels by dropping them into a spirit jar with a small quantity of a degreasing agent such as cigarette lighter fuel (although proprietary degreasing agent is available through material houses, I have used lighter fuel for many years and found it most effective). If they don't come apart easily by agitating the fluid, feed a pricker into the hole in the jewel and, holding the setting of the jewel with your tweezers, pull the setting and the jewel up over the pricker. You can do all this with the jewels in the fluid.

There are two reasons for separating each cap jewel from its jewel hole in this way: firstly, they don't always separate when put into the cleaning machine assembled; and secondly, we will be putting the jewel holes in one mini basket and the cap jewels in another to avoid a cap jewel and jewel hole coming together again in the machine, as they are often inclined to do, which prevents effective cleaning and drying.

Alternatively, you could put the jewel holes through the cleaning machine and clean the cap jewels by hand. To clean them by hand, remove one of the cap jewels from the degreasing agent, place the cap jewel on a piece of tissue paper, flat

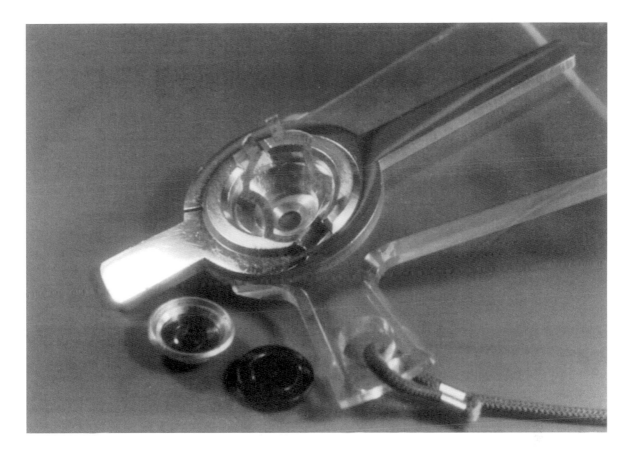

Fig. 7 Model of an Incabloc shock system with jewel hole, cap jewel and lyre spring.

side down, fold the tissue paper over the jewel, and move it around with the tip of your finger to remove any congealed oil. You will need to hold the tissue paper flat down to the bench, otherwise it will act as a springboard for the cap jewel. Next, carefully unfold the tissue paper and take up the cap jewel in your tweezers; hold it at an angle so that it catches the reflection of your bench light, and inspect its flat side for congealed oil and possible pits worn by the end of the balance staff – if it is pitted it will need replacing. If all is well, put it to one side for oiling. Now repeat this with the other cap jewel.

After the watch has been cleaned, remove the jewel holes and inspect these for congealed oil too. It is unlikely that you will find any other fault. If a jewel hole needs pegging, hold it down

on the bench with your tweezers while you peg in and around the hole from both sides. A useful technique to sharpen a piece of pegwood so that it will enter the hole in the jewel is to use a sharp craft knife while supporting the pegwood on clean tissue paper. This supports the tip of the pegwood, preventing it from breaking off, and also keeps it clean. (Fig.8)

Reassembly
Mostly the top and bottom jewel holes and their settings are identical, but there are exceptions, and where they are different, just remember which goes where. The top and bottom cap jewels may also well be different: this is more likely to be in their thickness (height), but it may be in their diameter too. If the only difference is

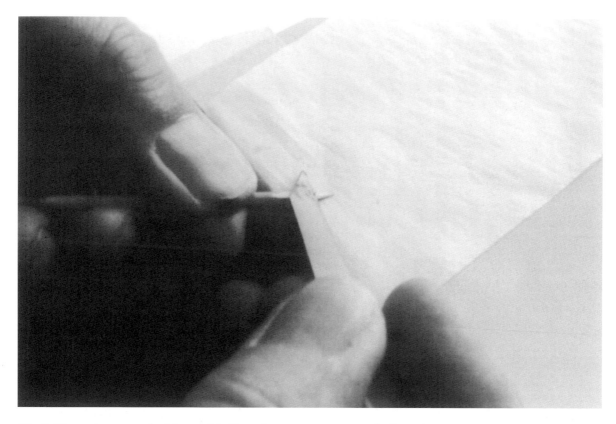

Fig. 8 Sharpening pegwood with a craft knife on tissue paper to preserve the fine point.

in the thickness, the thicker cap jewel is always associated with the balance cock; the thinner, after being re-united with its jewel hole, is fitted to the bottom plate. When there are differences in diameter, simply return each jewel to the more appropriate setting. Cap jewels are always a close fit in the jewel hole setting.

1. Having inspected both jewel holes and cap jewels, hold one of the cap jewels down to the bench, flat side up, and deposit a small quantity of Moebius 9010 oil to the centre of the jewel. The spot of oil should cover about two-thirds of the area of the jewel and have some height. (Fig.9)

2. Place the jewel hole and its setting over the cap jewel, making sure that the setting is held flat (Fig.10). Notice in Fig.10 the index finger of the left hand steadying the tweezers; I find this essential for success. If events don't go according to plan and the oil spreads to unwanted areas, just wash the offending parts in your degreasing agent and start again.

3. Turn the complete unit over and look for a good oil circle: ideally it should cover two-thirds of the area of the jewel, with the inside of the circle looking particularly crisp and clear, and the circle itself looking a bit like the outside of the bubble of a spirit level. (Fig.11)

Fig. 9 The cap jewel oiled ready for assembly.

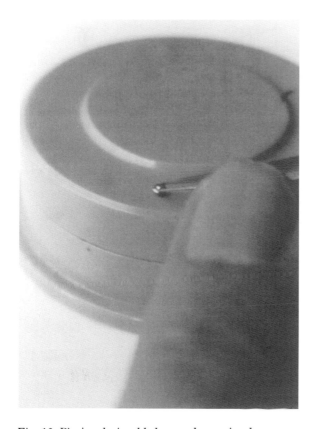

Fig. 10 Placing the jewel hole over the cap jewel.

After removing the bottom plate and balance cock from the cleaning basket, unlock the lyre spring again, replace the complete setting, and secure the lyre spring once more.

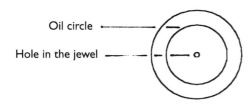

Fig. 11 A good oil circle covers two-thirds of the area of the jewel.

Watch Screws

For the experienced worker, most of the time, and certainly when working on simple fifteen-jewel lever watches, watch screws are fairly easily identified if they get mixed up in the cleaning basket. But when the watch has complications such as automatic work, calendar work or chronograph, even the experienced worker may be uncertain about which screw goes where. The difference between screws can be very subtle, yet it is vitally important for the functioning of the watch that they are in their correct position. It is quite common to find a small dent in the barrel of an automatic watch indicating that someone has used too long a screw to hold the automatic work onto the top plate, and that the end of the screw has butted the barrel. Sometimes the difference between two screws can simply be that the end of the screw has been rounded and polished because it can be seen from the back of the watch.

Even as an experienced worker, if I feel there is a chance that I might get similar screws mixed up, during the dismantling and cleaning of a watch I will return the screws to their correct place; then after cleaning, I remove them as necessary to replace components. I apply this method to chronographs in particular, as screws that were once identical may have been modified to compensate for wear – this may have involved removing as little as a couple of hundredths of a millimetre from under the head of a shouldered screw to prevent its associated part being pinched and to give it minimal endshake.

Alternatively you could draw circles on a sheet of paper, then label each one, and place each screw or component in its corresponding circle; then clean them individually by hand.

Try not to lose a screw, because named screws for a particular calibre are not generally available from material houses (except perhaps in twenties). Assortments of screws are easily obtained, but an exact replacement for a lost or damaged screw is often not possible, and so modification to the head or length of the screw may be necessary. Taps of watch screws are almost never altered, though on occasions a steel screw must be allowed to cut its own thread in a brass plate.

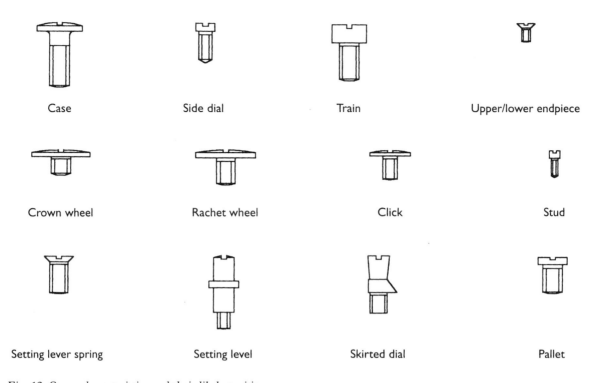

Case	Side dial	Train	Upper/lower endpiece
Crown wheel	Rachet wheel	Click	Stud
Setting lever spring	Setting level	Skirted dial	Pallet

Fig. 12 Screw characteristics and their likely position.

See Fig.12 for a list of screws together with their main characteristics.

Screwdrivers

If you have good quality screwdrivers, both the blade and the handle will differ in diameter. This will automatically help to give a screw the right torque, in the same way that the length of a spanner, when used normally, helps to give a nut the right torque. (Fig.13)

Keeping a good edge on a screwdriver is very important if it is to avoid slipping, and if it is to be efficient at removing screws. Personally I like to keep the blades of my watch screwdrivers in good condition by stoning on an India stone.

The width of screw slots do not seem to follow any particular rule with respect to a particular diameter head, so no strict rule can be given for proportions. My own watch screwdrivers range from 0.6mm in diameter to 3mm. The width of the ends of the screwdrivers varies between 0.1mm for the smallest, and 0.2mm for the

Fig. 13 Different diameter screwdrivers help give screws the right torque.

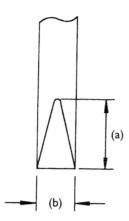

Fig. 14 The appropriate angle of a watch screwdriver is when (a) equals 1½ to 2 times (b).

largest. The stoned angle of the end of the blades is about 24°, which I achieve by ensuring that (a) is 1½ to 2 times (b). (Fig. 14)

Watch screwdrivers have a rotating top which may be round or hexagonal to give them an anti-roll quality. The screwdriver is held as shown in Fig.15.

What is a Watch?

This question is often asked, and I shall attempt to answer it in a way that will give the following chapters more meaning. The more terse amongst us might simply describe a watch as a form of time-measuring device that is portable and normally worn about the person. Horologists think of it as having two functions, firstly to tell the time, secondly to be decorative. As a rule, the question relates more to any time-measuring device, rather than to a watch specifically. The earth itself and the apparent movement of the stars and sun around the earth is frequently used by all of us for measuring time — for example, on a clear day we can tell the approximate time of day by the height of the sun above the horizon. Successive days are marked by sunrise and sunset.

In the past, time has been measured by monitoring the rate at which water drains from a vessel, and by burning candles, to name just two methods. However, there are several problems associated with this sort of time-measuring device, in particular that it can only ever be approximate. No doubt if we were going to use candles for telling the time, we would make each the same length and the same diameter – but even if the ingredients *were* the same, it is unlikely that two candles would take exactly the same time to burn, and if we wanted to measure longer periods we would have to be around to replace them as they burned out.

Our predecessors came to the conclusion that what was really needed to measure time's progress was a constantly recurring event which followed some natural law of physics, and which could be counted up. A pendulum follows the laws of gravity, but obviously cannot be used for a watch because it is not portable — so what to use?

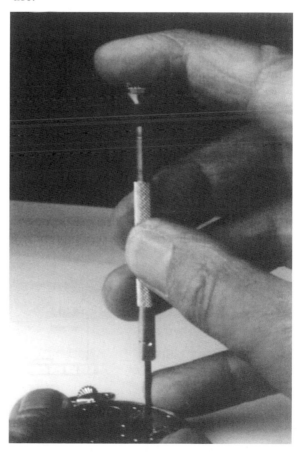

Fig. 15 The correct hold on a watch screwdriver.

Fig. 16 Represents one vibration of 540°.

In mechanical watches the measuring device is a type of wheel called a balance, which has a spring attached to its axis that is coiled in the form of an Archimedean spiral. The inner part of the spring is attached to a split cylinder of brass called a 'collet' that is fitted to the axis of the balance (the balance staff), while the outer end of the balance spring is attached to a post called a 'stud' which is held captive by a balance cock.

When the balance is stationary, it is said to be in the position of rest and therefore the balance spring will be unstressed and in a state of equilibrium. If the balance is now turned and released, it will start to vibrate first in one direction and then in the other; but each arc will be slightly less than the previous one, primarily because of frictional losses of the balance pivots in their holes, losses due to the internal friction in the balance spring and air resistance. However, in this arrangement we have a motion which can be repeated and counted, and it has great potential because it behaves according to the laws that any spring will follow. Arcs of the same angle will take virtually the same time to complete, and even the time taken for differing arcs will be acceptable for practical purposes. However, we still have other practical problems to overcome.

Maintaining the Action of the Balance

In the example given above, the balance was given a slight turn to start it vibrating. To maintain the vibrations, the balance is given a push during each arc to maintain the vibrations: this push is called an impulse. The strength of the impulse will be sufficient to maintain arcs of vibration of between an absolute minimum of 440° and up to approximately 640°. (Any higher arc of vibration risks the impulse pin striking the wrong side of the notch as the watch is subjected to strong movements.)

A vibration can be defined as the arc a balance turns through by swinging from maximum displacement on one side of the position of rest, through the position of rest, to maximum displacement on the other side of the position of rest (see Fig. 16).

The Power Source

The driving or motive force for a mechanical watch is a mainspring: in theory this is a straight ribbon of steel coiled into a barrel and wound even tighter so that with one end of the mainspring anchored to the axis of the barrel (the barrel arbor) and the other end locked into the wall, a turning force is put onto the barrel.

Winding Mechanism

When the stem is turned, it turns the clutch wheel which drives the winding pinion. This drives the crown wheel, which in turn drives the ratchet wheel, which is secured to a square on the barrel arbor and is prevented from turning

Fig. 17 The winding mechanism

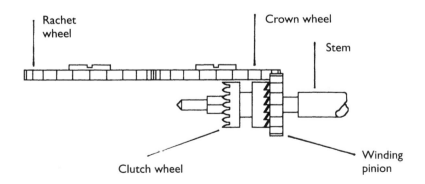

Rachet wheel

Crown wheel

Stem

Clutch wheel

Winding pinion

backwards by a pawl arrangement called a click (no doubt because of the sound it makes while winding). (Fig 17)

Transmission of Power

Power from the mainspring is passed to the balance through a series of wheels with a particular ratio between them, so that rather than count up the number of vibrations of the balance, which is what really happens, one of the wheels turns once an hour and so carries an indicator called a minute hand that measures an hour and parts of hours. Collectively these ratio wheels are called a train, and give a step up in gear ratio throughout the train.

The train most commonly consists of a barrel; a centre wheel which turns once an hour; a third wheel; and a fourth wheel which may have an extended pivot to carry a seconds hand. The running down of these four wheels alone is a constantly occurring motion and so could be used to measure time, but one problem would be the frequency with which the arrangement would need rewinding (at least once a minute). Moreover, each unit of time would vary slightly from the last, and this would accumulate to unacceptable errors. An arrangement was required that would control and slow down the release of power so that winding need only take place once a day. Such an arrangement is effected through what is known as an escapement.

The Escapement

The fourth wheel of the train drives an escape wheel which gives impulse to the balance through a pallet. The combination of escape wheel (often abbreviated to 'scape' wheel), pallet and balance is called the 'escapement', effectively a geometrical arrangement to keep the balance vibrating. The balance, of course, controls the speed with which the power is released, so scape wheel and balance are mutually dependent.

Regulation

The arrangement so far has significant potential, although it needs a mechanism for fine tuning or regulation. This can be achieved in two ways: by altering the weight of the balance – not necessarily by adding weight, but simply by moving mass from the rim of the balance towards the balance centre; and by altering the effective length of the balance spring. The former is sometimes done on high-grade watches, but the latter is far more common. A regulator is fitted at a point about a quarter-way back from where the outer end of the balance spring is pinned, the balance spring (sometimes called a 'hairspring' because of its fineness) passing between an index pin and boot in the regulator so that it can be turned, altering the effective length of the spring. Shortening a balance spring causes a balance to vibrate faster; conversely, lengthening a balance spring causes a balance to vibrate more slowly. (Fig. 18)

Framework

The barrel, train and escapement is held in a framework made up of a plate, cocks and bridges. For example, normally the train is placed into a bottom plate which is most often circular but could be other shapes. The barrel and centre wheel are often held in with a barrel bridge; the rest of the train, including the escape wheel, with a train bridge; and the pallet with a pallet cock. The balance is normally held with a balance cock. The difference between a cock and a bridge is that whereas a bridge is a plate with support on two sides, a cock is a plate with support on one side only. Confusingly the rules are not strict, and you often get cocks called bridges, and bridges called cocks.

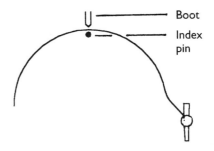

Fig. 18 Regulation is by shortening or lengthening the balance spring by moving the regulator.

Ratios

What we call 'wheels' usually have both a wheel and a pinion. The wheel is usually made of brass and has teeth, and the pinion is made of steel and has leaves. The number of teeth and leaves, together with the timing of each vibration of the balance, will enable the correct ratio to show the time of day, rather than the number of vibrations of the balance in a day. To achieve a 12:1 ratio between minute hand and hour hand, we have 'motion work' under the dial which includes a cannon pinion, a minute wheel and pinion, and an hour wheel.

Display

Remembering that our objective is to tell the time, we need a display. What we understand as the conventional display through hours, minutes and perhaps seconds is actually an analogue display – though watch repairers didn't use the term 'analogue' until quartz watches with alternative displays appeared.

Hand Setting

Of course our watch is unlikely to keep perfect time, and we may allow it to 'run down' occasionally – and so that we don't have to wind it at the time shown on the dial, a clutch arrangement is incorporated so that we can set the hands in an independent operation, ie. without having to turn all the other wheels in the clock simultaneously.

In Summary

To summarize then, a mechanical watch is essentially carried about our person; it is to look attractive, and be functional, which means keeping good time; it incorporates a motion which recurs and gets counted up; and it has a framework, a power source, a winding mechanism, a means for transmitting power, an escapement, a means of regulation, and finally a display and a facility for hand-setting.

The role of the professional repairer is to make sure that the watch continues to function and to look attractive, whilst achieving job satisfaction and making a living in the process. The enthusiast will derive joy from enlightenment, and the same satisfaction from seeing a watch that he has worked on, go well.

This is, of course, an oversimplification, and it is perhaps one of the attractions of watch repair work that you never actually arrive, you just get nearer your goal. For example, just the one aspect of adjusting balance springs is something that in repair work one can only get better at.

Maintaining mechanical watches is about combating the ingress of dirt and sometimes water, and the deterioration of lubricants; it means making good general wear and tear both in the case and the movement; and it also involves assessing the problems associated with positional errors, and to a lesser extent in modern horology, temperature changes. Any change in barometric pressure does affect watches, but there is nothing the repairer can do about it unless the watch can be made water-resistant: when we make a watch water-resistant to a particular pressure, indirectly we do ensure a constant air pressure inside the watch.

Lubrication Chart

It is not realistic for repairers to keep every make and every grade of lubricant recommended by every watch manufacturer. Most of the time, watch repairers simply transfer knowledge about lubrication from watch to watch based on whatever they have picked up over the years. I acknowledge that manufacturers have resources in considerable excess to my own, but in the course of my practice I have been able to select what I consider to be the most prestigious of brand names and have lubricated watches accordingly. The following table is a consensus of 'best practice', and I recommend it to you in the absence of specific recommendations from manufacturers. Equivalent lubricants by other manufacturers may also be used. (Jewelled impulse pins and jewelled pallet pivot holes are not oiled. If they are metal, oil very lightly with light watch oil such as Moebius 9010.)

Section	Lubrication Point	Lubricant
Barrel and mainspring	Mainspring (other than brand new) Arbor pivots in the barrel and cover Arbor pivots in the plate	Microtime watch grease
Centre wheel	Top and bottom pivots – cannon pinion tension	Moebius 9020
Click	Pivoting point and pressure point of click spring	Moebius 9020
Train	Third wheel top and bottom pivots Fourth wheel top and bottom pivots Centre seconds pinion pivoting points Tension spring for centre seconds	Moebius 9010
Pallet	Impulse face of both pallet stones (Jewelled pallet pivots are not oiled)	Moebius 941
Winding and handsetting	Crown wheel centre Winding stem, pivot, square, groove, main bearing and the thread close to the button Winding pinion, clutch teeth and outside face Clutch wheel, clutch teeth and groove Setting lever screw Setting lever where it engages the yoke Yoke pivoting point Setting wheel post Minute wheel post Setting lever spring where it engages the setting lever	Microtime watch grease

2 CLEANING A TYPICAL FIFTEEN-JEWEL LEVER WATCH

Dismantling, Cleaning, Reassembling, Oiling, Adjusting

PART NAMES AND NUMBERS

It is necessary to be able to identify watch parts for two reasons: firstly, to understand the following text; and secondly, so that correct parts can be supplied against an order to a supplier. Horologists in different English-speaking countries often use different terminology for the same part, and even two watch repairers in the same workshop may use different terminology; because of this, a universal system of part names and numbers has been devised. Throughout this book the Swiss preferred names and numbers have been used where possible; these are the original preferred numbers, as distinct from the more recent ISO numbers which have still not been universally adopted. Material houses seem to prefer the old, original numbers. (Fig. 19)

A comprehensive book of watch part names and numbers entitled *The Horological Dictionary* is available from some material houses and ETA in Switzerland; also two books on the interchangeability of parts between ETA calibres; and a book which identifies part names and numbers of mechanical and quartz watches.

TESTS BEFORE DISMANTLING

For the purposes of this book a typical fifteen-jewel lever watch calibre Unitas 6497 has been chosen as representative of other watches. Although the instruction given is specific to that calibre, additional comments are made to accommodate alternative movements so that what is learned here is transferable to other

calibres. Before you set out to overhaul a watch, you will first need to establish why it has stopped. The vast majority of watches stop because the old oil has dried and congealed, and perhaps the case has been letting dust in. However, the following examinations and tests should be made:

- Wind the watch to establish whether the mainspring has broken. If the spring is all right you will feel a resistance to winding when it is fully wound; if, however, the watch can be wound indefinitely and it is not an automatic watch, the mainspring is broken.
- While you are winding, listen out for the sound of the teeth slipping between the clutch wheel and winding pinion – an irregular clicking noise. If there is any sort of wear you will need to fit a new clutch wheel and winding pinion.
- Pull the stem out and push it in again to establish a positive action, and a positive position for the button (also known as 'crown') and stem. If the positions are uncertain, the setting lever spring will be broken and a new one should be fitted.
- Set the hands, and as you turn them, establish that there is a slight resistance: the centre wheel and cannon pinion have a clutch arrangement so that although the hands carry with the centre wheel, they may be turned independently for hand setting. If the hands move exceptionally easily, the cannon pinion is slack and will require tightening or perhaps even replacing (though this is rarely necessary).
- Check the condition of the case. A watch has two primary functions: the first is to tell the

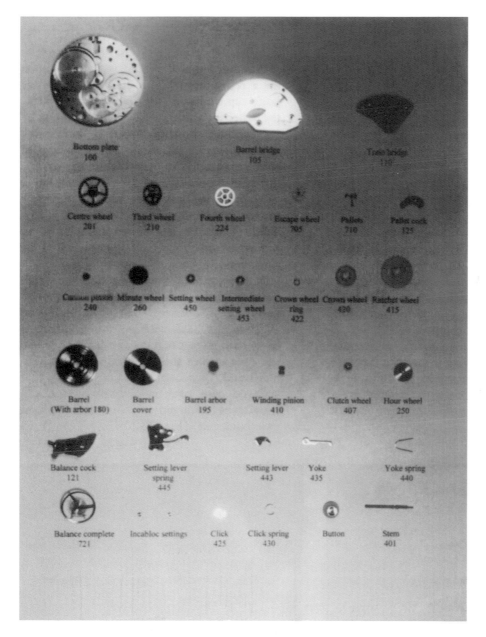

Fig. 19 Part names and numbers.

time reliably, the second to be pleasing in appearance. Little or nothing can be done to improve worn chrome or gold-plated cases, although it is possible to improve gold and stainless steel cases.

- Check that the glass is in good condition. If it is cracked it will need replacing; if it is scratched it could be replaced, but a realistic option is to polish an unbreakable glass with a very fine grade of emery cloth, then with Autosol on a cloth.
- Check that the button is in good condition, functionally and in appearance. If it isn't, replace it.
- Remove the case back (see below) and check that the balance staff is not broken.

Fig. 20 Unitas 6497 pocket watch.

If the watch passes all these tests yet loses badly or stops, it probably needs no more than a good clean and lubrication and just minor adjustments. Some repairers demagnetize mechanical watches before overhaul as a matter of course, others only do so if the watch is found to be magnetized. The first sign of magnetism is usually when steel parts stick to carbon steel tweezers.

DISMANTLING

1. Remove the case back: it may be screwed, or it may be a snap-on. If it is snap-on, look for a relief on the case back where a case knife can be inserted. With a pocket watch, the relief for the bezel is often at about the two o'clock position, and at about the eleven o'clock position for the case back. On a wrist watch the relief is most usually opposite the button. Insert the case knife, then press and tilt the back of the knife upwards; this prevents it slipping (Fig. 21). There may be an inner case back to remove as well, for which you proceed in the same way.

    If the case back is screwed, hold the watch between the palms of the hands, and press and twist to unscrew. Be careful if you unscrew the back with your fingertips: the back may stick causing your fingers to slip and, because of the sharp edge on the back and bezel, they could get cut. (Fig. 22.) A screwed case back on a wrist watch can usually be removed with a water-resistant case-back tool. (Fig. 23)

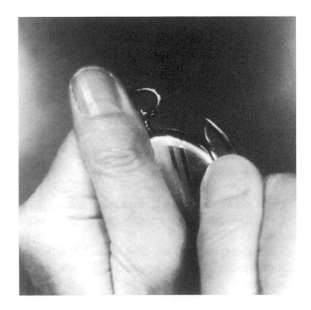

Fig. 21 Removing a snap-on case back.

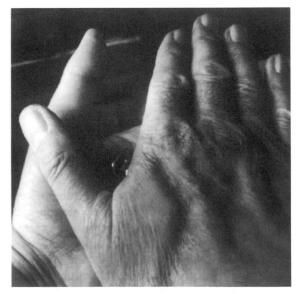

Fig. 22 Removing a screw back on a pocket watch.

Fig. 23 Water-resistant case backs are removed with one of a variety of water-resistant case-back openers.

5. Turn the hands so that they are on top of one another (as at a quarter to nine for example), cover them with tissue paper, insert hand lifters, and lift them. The tissue is to protect the dial from the lifters, and to stop the hands from 'pinging' (flying away). (Fig. 25.) Place the hands in the material tray. Do not remove the small seconds hand at this stage: it comes off with the dial.

6. Slacken the setting lever screw by about two turns, and pull the button and stem from the watch and case. The setting lever screw is the small screw near where the stem enters the movement. After slackening the screw, it may be necessary to push lightly on the screw to assist the setting lever to clear the groove in the stem. Some wrist watches have a setting lever that pushes forwards instead of having a screw.

7. Remove the two case screws. Place them in the cleaning basket.

2. If your practice is to allocate a job number to the work you will be doing on the watch, inscribe this number into the back of the case. If you keep records of all the work you do, this may be useful in tracking down what you last did to the watch should it be returned to you in the future for whatever reason.

3. Let the power off the mainspring by winding slightly, disengaging the click with a piece of pegwood and letting the button turn under control between your index finger and thumb. (Fig. 24) It is advisable to stop the balance from vibrating as the power comes off so as to avoid damaging the scape wheel, pallets or balance.

4. Remove the bezel in the same way as the back was removed.

Fig. 24 Letting the power off the mainspring.

Fig. 25 Lifting the hands with tissue paper and hand-lifters.

When putting parts into the cleaning basket, take great care that large, heavy parts do not damage smaller, delicate parts during the cleaning operation. Also, try to avoid mixing screws and having to sort them out later. To this end, I usually put the screw and its associated part together. For example, I wouldn't put the train bridge with the train, but I *would* put the train bridge screws with the train. Screws are light and unlikely to cause damage to the train. Fig. 26 shows how you might lay out a three-tier cleaning basket when cleaning only one watch. In fact I have used two tiers as the watch in question is a pocket watch and too large to fit into one tier alone. As an experienced worker, working on wrist watches, I usually have three identical sectioned tiers with at least one watch in each, but sometimes with a total of five watches in three tiers; it depends on the size of the watch and the degree of complication. Here we are sticking to one watch spread over two tiers; the middle tier will remain empty.

8. Push the movement out of the case.
9. Put the case parts to one side for cleaning later.

Next we must remove the dial, but before we do so we will first consider how dials are secured. This is generally in one of three different ways:

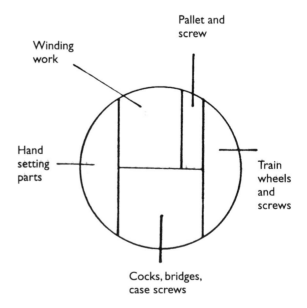

Fig. 26 Suggested layout for the cleaning basket.

(a) By skirted dial screw.

(b) By side dial screw.

(c) By a 'dial fastener' that swings on a post and holds the dial to the movement in a similar fashion to the side dial screw.

Each type has two or three copper feet hard-soldered to the back of the dial. With the skirted dial screw, before a new dial is put on the movement, the dial screws are inserted into the bottom plate, tightened, then backed off until the hollow, or flat, in the screw is adjacent to the hole in the plate for the dial foot. The dial is then put on and held tightly against the bottom plate, and the screw is unscrewed by about a quarter of a turn so that its skirt cuts into the copper foot and pulls the dial to the bottom plate. To remove the dial, screw up the dial screws until the hollows or flats are adjacent to the dial feet, then simply lift the dial off.

Side dial screws have pointed ends and screw into the edge of the bottom plate; after the dial is replaced, they are tightened against the dial feet.

Modern dial fasteners swing horizontally to cut into the copper dial feet after the dial has been replaced. (Fig 27)

Coming back to our demonstration watch:

10. Slacken the side dial screws by one or two turns and gently prise the dial away from the bottom plate, taking care to prevent the seconds hand from flying away. Retighten the screws after removing the dial to prevent them coming out in the cleaning machine.

11. Remove the dial washer, if fitted, and the hour wheel. Place them in the cleaning basket.

12. Remove both Incabloc settings – one in the balance cock, the other in the bottom plate – and put them in the spirit jar as suggested on p. 10. Remember to lock down the Incabloc springs in the plate and cock.

13. Feel the play on the barrel to confirm that all the power is off. Do this by resting pegwood between two barrel teeth and rocking the barrel clockwise and anticlockwise. It should move freely.

14. Put the movement into a movement holder with the top or back of the movement facing up.

15. Remove the balance cock screw.

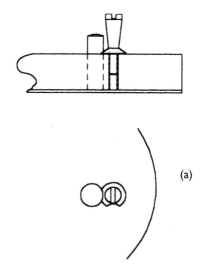

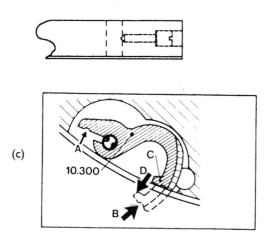

Fig. 27 (a) Skirted dial screw. (b) Side dial screw. (c) Dial fastener. Open by pushing at 'A' or pulling at 'D', close by pushing at 'B'.

16. Remove the balance cock by lifting the cock, turning and agitating it to free the impulse pin from the notch, then lifting the balance free. The balance spring will easily support the weight of the balance. Rest the balance and cock on the bench.

17. Turn the cock over, then turn the boot by 90° for the balance spring to pass freely between the index pin and boot. If the boot doesn't want to turn and there is a risk of shearing it off, simply bend the index pin away from the boot.

18. Hold the cock in your left hand while you slacken the stud screw, and push out the stud with the point of your tweezers; then retighten the stud screw to prevent it from coming out altogether and being lost. Put the balance in your material tray and the balance cock in the cleaning machine.

Modern practice is to fit the stud into the stud holder by friction rather than by using a screw. In some instances the stud holder is still provided with a screw hole, although the hole is not used. If there is any problem at all in removing the balance from the cock – for example, if it is fitted with the new system of stud and regulator known as Etachron (distinguishable by a split regulator with a combined index pin and boot that can be rotated and a split stud holder) – the balance and cock may be replaced on the bottom plate without the Incabloc unit and cleaned in the cleaning machine as it is. In this case, the cock is removed simply to gain access to the pallets and pallet cock, and is then replaced again.

19. Remove the pallet cock screw, the pallet cock and the pallet. If the bottom pallet pivot is tight in its hole, push the pivot out from the other side rather than pull on the pallet. Pulling the pallet will almost certainly lead to a broken pivot.

20. Remove both the crown wheel screw, which is left-hand thread, and the ratchet wheel screw which is almost always right-hand thread. (It should be noted that AS calibres

often use a left-hand thread for both screws.) Any watch screw with three slots in the head will be a left-hand thread, though many left-hand screws are not marked in such a way. If you are unsure of the direction to turn a screw, build up the pressure slowly in each direction until the screw gives in one direction.

21. Remove the crown wheel and the crown wheel ring.

22. Remove the ratchet wheel.

23. Remove the screws holding the train bridge and the barrel bridge.

24. Remove the train bridge, the third wheel, fourth wheel and escape wheel. Put them in for cleaning, but remember to separate the bridge from the wheels.

25. Remove the barrel bridge. It is perfectly in order to remove the click and click spring, but if there are no problems associated with these, they may be left – this is considered by many repairers to be a legitimate short cut. If you are unhappy with this, remove them, but if you do, be careful not to lose the click spring – and don't put it in the cleaning machine because it will almost certainly fall through the mesh: clean it in the spirit jar. To remove a shepherd's crook-type click spring, hold the crook of the spring with the fingernail of the index finger while unloading the spring with the tweezers.

26. Remove the clutch wheel and winding pinion.

27. Remove the barrel for cleaning.

Since the advent of unbreakable mainsprings, and because such mainsprings are sold as unbreakable, rustless, antimagnetic, non-setting and self-lubricating, some manufacturers advocate not stripping the barrel for cleaning but to brush through the barrel teeth, peg around the barrel holes, and lubricate the barrel arbor pivots before replacing the barrel. Personally that practice doesn't rest easily with me, consequently I prefer to strip the barrel, taking the cover off at least and removing the barrel arbor.

28. Mark both barrel cover and barrel wall with a suitable tool (a scribe will do fine) by just pricking both so that they can be reassembled in the same relative position.

29. To remove the cover, hold the barrel, with the cover facing down, between the thumb and the first two fingers, then strike the barrel arbor with a brass hammer or watch brush; the cover should spring off. Alternatively, rest the barrel on the bench with the cover facing up, push on opposite sides of the barrel on the teeth, and the cover and barrel should separate.

30. Carefully lift off the cover. Turn the arbor with the tweezers against the pull of the mainspring, keeping your finger in the spring to prevent it being pulled out of the barrel, then lift the arbor out.

Some repairers will carefully remove the mainspring from the barrel; others will put the barrel into the cleaning basket without removing the mainspring – though in this case be sure to have the spring facing down to allow the cleaning and rinsing fluids to drain out. Barrels for automatic watches are treated differently, and their treatment is explained on p. 107.

31. Remove the bottom plate from the movement holder.

32. Remove the cannon pinion with a suitable tool. An old pair of top cutters with the cutting edges blunted on a stone to prevent marking the cannon pinion is acceptable for this, though special cannon pinion-removing tools are available. (Fig. 28.)

33. Remove the centre wheel and examine the top and bottom pivots. If they are cut it will probably make more sense to obtain a new centre wheel than to attempt to polish the worn pivot and have to bush the plate or bridge. If a new wheel is unobtainable, polishing and bushing may be the only alternative.

34. Slacken the setting-lever spring screw, unload the setting-lever spring from the setting lever, and continue to unscrew the screw; remove this and the spring, and put them in the cleaning basket. It is quite common to find the setting lever spring broken. If it is, it should be replaced, otherwise the button and stem may not be held in the hand-set position safely, and the stem may return to the wind position during hand-setting, much to the annoyance of the owner.

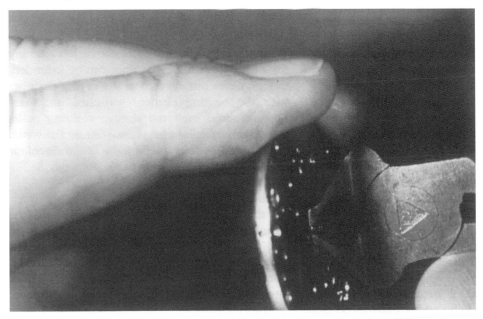

Fig. 28 Cannon pinion removal.

35. Remove the minute wheel, the setting wheel and the intermediate setting wheel (though usually there is only one setting wheel).

36. Prevent the yoke spring from flying by placing a fingernail over the crook in the spring while you hold the long end of the spring in your tweezers and remove its tension. Lift the spring and put it in the material tray for cleaning in a spirit jar later. Don't put it in the cleaning machine or it will probably fall through the mesh.

37. Remove the yoke and put it in the cleaning basket.

38. Unscrew the setting-lever screw and remove both the setting lever and the screw. This is not essential, but is useful practice on your first watch just to familiarize you with the way things are.

Notice that little has been examined as we dismantled. This is because the faults we would be looking for, such as faulty pivots and jewel holes, are too subtle to show up until after cleaning. Other faults – a broken balance staff, for example – would have become apparent without looking too closely.

39. Load the basket frame, put the lid on and put the watch through the cleaning machine.

While the watch is cleaning, it is usual to degrease the balance; if it is not put through the cleaning machine, clean the case and clean any shock-proofing parts, and degrease the balance itself with lighter fuel in a small spirit jar. Finally dry with absorbent paper towel; this should remove any residual oil picked up from in the degreasing agent. Personally I would clean a maximum of five balances in degreasing agent before I disposed of the fluid, cleaned out the spirit jar and put in fresh fluid.

WATCH CLEANING

Virtually all cleaning of watches is carried out using a cleaning machine and usually involves a cleaning cycle, two rinsing cycles and one of drying in a pre-heated chamber. Cleaning is effected by the chemical action of the cleaning fluid, together with its mechanical action as it swirls around the various parts. These two actions may be assisted by a vibration unit, which is part of the framework for holding movements, or by ultrasonic waves generated in the cleaning machine.

Watch-Cleaning Machines

The most basic of machines for commercial use, but still effective, are manually operated with a variable speed; they have one cleaning jar and two jars for rinsing – but usually the same fluid is used in each – and a drying chamber with a switched heater element. Sometimes the cleaning machine has a timer to remind the repairer that a cycle is completed.

Although timings can vary, over a number of years with different cleaners and fluids I have found the following timings about right, with half a minute spin-off between cycles:

Cleaning	2½ min
First rinse	1½ min
Second rinse	1½ min
Drying	4 min (heater element switched off)

When a watch is very dirty, it can be left cleaning for a little longer. Drying time can be extended, though never put a watch in the drying chamber with the heating element switched on: over-heating could melt the shellac securing pallet stones and impulse pins, it may cause the watch to rust, and serious overheating could discolour and soften steel and brass parts.

Automatic Cleaning Machines

As the name suggests, you simply load the cleaning basket into the machine, switch it on and it automatically runs through the various cycles of cleaning, rinsing, spin off and drying. Some automatic cleaning machines offer alternative programmes such as variable speed, continuous rotation, and reciprocating action, in which the basket rotates in alternate directions at variable fixed speeds and may have a vibration attachment with an eccentric weight.

Ultrasonic Cleaning

Personally I feel that the ultrasonic cleaner is the most efficient type of cleaner I have used. It works through a transducer that generates ultrasonic waves, which in turn cause imploding bubbles to bombard watch parts during the cleaning and rinsing cycles. In the 1960s some watch repairers who cleaned ultrasonically complained of loosened pallet stones and impulse pins, but this has not been my experience. The strongest criticism I have is that any hairs present in the cleaning or rinsing fluids can find their way to the bottom of blind holes – but even that happens very rarely, besides which it is not a problem as they cling together and are easily removed. To my mind ultrasonic cleaning is very effective, and almost indispensable for cleaning cases and watch bracelets.

When cleaning watch bracelets and cases, I use a separate ultrasonic tank that has wire mesh baskets that can be loaded and placed into the ultrasonic tank at will.

Cleaning and Rinsing Fluids

There is a wide choice of cleaning and rinsing fluids. For me they must be effective, they should preferably suit both watch and clock work, and must be as non-toxic as possible; also the cleaning and the rinsing fluids need to be compatible, they should be available through my normal material supplier, and they must suit the method of cleaning – for example if cleaning ultrasonically, I would choose fluids known to be suitable for this type of cleaning.

Cleaning fluids include ammoniated, non-ammoniated, water-based, waterless, non-foaming, those that suit both ordinary and ultrasonic cleaning machines, ready mixed, and concentrates to be mixed with water. Concentrates have the advantage of ease of supply and storage but cannot be used for any technique involving partial dismantling because of the increased risk of rusting.

Rinsing fluids must be fast to dry, leave no spotting and be compatible with the cleaning fluid. There is a rinse to suit all cleaners, and some are as suitable for ordinary cleaning as they are for ultrasonic cleaning. I never put 'plastic' parts into conventional cleaning or rinsing fluids; with new materials being developed all the time, compatibility is often unknown. Instead, I wipe plastic parts over with isopropyl in the form of injection swabs sometimes used in the jewellery trade when ear piercing.

In addition to the above, there are special cleaning and rinsing fluids and special movement holders for cleaning partially dismantled watches. Perhaps this is not the place to discuss the ethics or desirability of partial dismantling (sometimes referred to as non-strip cleaning), but in my view the technique certainly doesn't suit all watches all of the time. I feel that those new to watch repairing would need considerable experience with complete dismantling before being able to make valid judgements as to the merits of partial dismantling and where it might be appropriate.

Jewellery cleaning fluids should be considered for case and bracelet cleaning, though I have also had good results cleaning ultrasonically using liquid hand soap with hot water as a rinse. Thorough drying is a must if black stains on shirt cuffs are to be avoided.

A Cautionary Note

Consider very carefully before following some of the older traditional recommendations for cleaning, rinsing and degreasing agents. More is known about the harmful effects of some of the fluids that used to be used, and I recommend that you ask about the safety aspect before deciding whether to use them. Benzene is one that immediately comes to mind: this used to be used as both a rinse in the cleaning machine and a degreasing agent; however, it can be particularly harmful, and I no longer use it in the workshop.

Using cleaning, rinsing and degreasing agents:
1. Follow the recommendations of the manufacturer.

2. Avoid using in a confined space.

3. Ensure adequate ventilation through an extractor fan or window.

4. Don't smoke or use a flame near the fluids.

5. Don't store large quantities of fluid in the workshop.

6. Don't tip used fluids down the sink or drain. Label containers of waste fluid stating the nature and quantity of the fluid, and dispose of them through your local authority.

7. Label all storage containers.

8. Know what to do in the case of an emergency.

Look for sediment at the bottom of the cleaning jar after cleaning five to ten watches. Even if a small amount can be seen, tip the fluid into a clean spare jar, throw away the sediment and top up with clean fluid and clean the jar. Check the first rinse and do similarly but top up from the fluid in the second rinse. After decanting the final rinse, top up with fresh fluid.

Case Cleaning

There are two commonly acceptable ways of cleaning watch cases and bracelets: (a) by scrubbing with a brush, such as you might use for dish washing, in warm soapy water and drying thoroughly; and (b) by ultrasonic cleaning machine. I find little to choose between the two in terms of appearance and time, but whilst the former is inexpensive, effective and pleasing on watch cases, it is not very effective at cleaning watch bracelets. By comparison the latter is expensive in capital outlay, but it is very effective on both watch cases *and* bracelets. Probably the enthusiast will justify the former, whilst the full-time professional repairer will find the latter economical in the long term. It is not essential to remove the glass for either method.

Do *not* be tempted to clean the case by digging at the dirt with pegwood. It scratches the case and is far less effective and less satisfying than the other methods described here – nor is it any quicker, either.

REASSEMBLY OF THE MOVEMENT

1. Replace the Incabloc units into the cock and bottom plate. If there is a different thickness between the two cap jewels, put the thicker one in the balance cock.

2. From now on, hold parts in tissue paper, and not in bare fingers. Fingerprints on watches are not tolerated.

3. Examine the jewel holes in the bottom plate, particularly the flat side. You are looking for cracked jewels, chipped holes and congealed oil. Cracked and chipped holes will need replacement jewels (see *The Clock Repairer's Manual* by the same author), congealed oil should be pegged off until the jewel is immaculate; nothing less. Again, the technique for examining is to get the bench lamp reflected in the jewels. Faults show up well under these conditions.

4. Put the bottom plate in the movement holder.

Now for a brief word on the function of the setting lever and screw. The setting lever screw is positioned between the plates with minimal endshake and sideshake; on its bottom end is the setting lever which holds the stem in position (Fig. 29). When the setting lever screw is unscrewed, instead of the screw rising, the setting lever moves away from the plate and its end comes out of the groove in the stem so the stem may be pulled out.

5. With Microtime watch grease, grease and replace the setting lever screw – though don't grease the thread. It is, however, important to lubricate the setting lever screw, as removing the movement from the case depends on being able to slacken the screw – besides which this is one of the first places to be attacked by rust.

6. Inspect the barrel, particularly the teeth; they must be upright, with no grit lodged between them.

7. Grease the mainspring if it is not being replaced with a new one. Although manu-

Fig. 29 Shows how the stem is held by the setting lever.

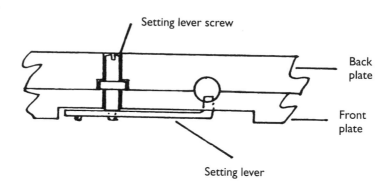

facturers of modern unbreakable mainsprings claim they are self-lubricating, every practitioner I have asked does grease them. Certainly if the mainspring is steel, it must be greased. Steel mainsprings are distinguishable by their blue colour; unbreakable ones are normally white in colour, similar to polished steel.

8. Inspect the barrel arbor. Grease the pivot that works in the barrel, and replace. As you replace the barrel arbor, turn the arbor in the opposite direction to winding so that the inner coil expands slightly. Engage the barrel hook with the eye in the mainspring.

9. Grease the other pivot which engages with the barrel cover, and replace the barrel cover, aligning the previously made marks. Clip the cover on using tweezers and fingernails as vices.

10. Feel the endshake of the arbor in the barrel – about 0.03 to 0.05mm would be safe. Certainly a slight up and down movement must be felt.

11. Grease the pivots of the barrel arbor that engage with the plates, and replace the barrel arbor into the bottom plate.

12. Inspect the centre wheel teeth, pinion leaves and pivots. Rust or cuts in the pinion leaves will necessitate a new centre wheel, as will cut pivots.

13. Oil the centre wheel front pivot with Moebius 9020, and replace in the watch.

14. Inspect the third, fourth and scape wheels, and replace. Check particularly that pivots

are straight and not cut, that there are no cuts or rust in the pinion leaves and check the teeth of the escape wheel. The locking corners of the teeth must be sharp and not rounded through wear.

15. Inspect the barrel bridge, and replace.

Most checks on watches are quite specific, but others are just general, with the watch repairer attending to whatever presents itself. For example, I wouldn't look specifically for a missing steady pin in a cock or bridge, yet I would notice if one was missing and would make a new one from brass pin wire.

16. Usually there are three barrel bridge screws, one of which is frequently – but not always – shorter than the other two. If the setting lever screw is to the left of the stem, viewed from the back of the watch, then the position for the short screw will be the nearest screw hole to the left of the setting lever screw. If the setting lever screw is to the right of the stem, then the short bridge screw will be the nearest screw hole to the right of the setting lever screw. The reason for this is so that the bridge screw will clear both the setting lever, and also hand-setting work (Fig. 30).

17. Inspect the train bridge jewels for cracks, chips around the hole and congealed oil.

18. Replace the train bridge. Place the thumb of the left hand under the movement holder, and rest your fingers lightly on the train bridge with a piece of tissue paper between

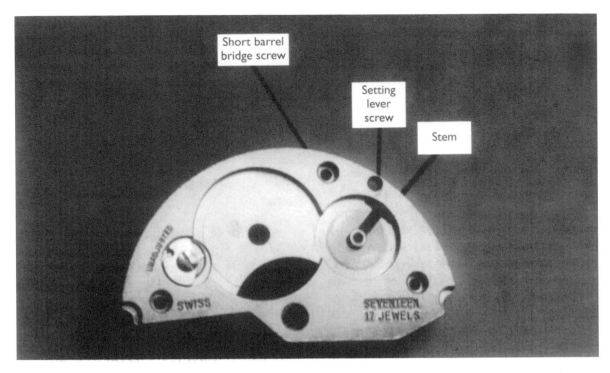

Fig. 30 The setting-lever screw lies between the stem and the short barrel bridge screw.

your fingers and the plate. With a clean watch oiler, manipulate the pivots into their holes. I prefer a clean oiler to tweezers due to the flexibility of the oiler. It can reach into the movement better than tweezers, and it will flex rather than break a pivot on the odd occasion when a wheel resists being moved as a pivot is coaxed into its hole.

19. Place the movement back on the bench holding the train bridge in position, and replace the train bridge screws. Tack down the train bridge screws lightly at first in case a pivot has come out of its hole. With your screwdriver, turn the centre wheel just a little to check that everything is free. Now tighten the train bridge screws.

20. Feel the actual endshake of each individual train wheel by lifting each wheel in turn and checking that it drops under its own weight.

21. Replace the ratchet wheel. If there is a recess in the centre of the wheel it will face up

usually as it is recessed for the screw. To replace the ratchet wheel, position it over the square of the barrel arbor, with its square just a little anticlockwise of its correct position. Now, with your fingernail on one side of the ratchet wheel, turn the ratchet wheel clockwise with your tweezers, allowing it to locate on the barrel arbor square and until the click engages with the teeth. Replace the ratchet wheel screw. A fingernail in the teeth of the ratchet wheel will hold it while the screw is tightened. Use tissue paper between your finger and the wheel. If a finger mark is left on the watch, it can be removed easily with Rodico, purpose made for removing finger marks and oil smears.

22. Now we are going to check again the freedom of the train. With your screwdriver, turn the ratchet wheel clockwise by half a turn. Hold the ratchet wheel lightly against the click, (lightly so that you don't unscrew

the screw again), and observe the train. It should be seen to run down and the escape wheel should, after running down freely, reverse slightly. This is evidence that the train is free. If it doesn't reverse, and not all watches do even when there isn't a problem, investigate just to make sure nothing is wrong. A free train, before lubrication, usually does reverse slightly after running down.

23. Replace the crown wheel ring, grease the outside, replace the crown wheel, grease the recess for the screw, and replace the screw. Remember it has a left-hand thread.
24. Check that the crown wheel is free.
25. With Moebius 9020, lubricate under the screw for the click; capillary attraction will pull the oil to its pivoting point. Lubricate where the click spring engages with the click.
26. Inspect the pallet pivots, stones, notch and guard pin. The pivots must be polished and straight; the stones must be secure, with sharp corners, clean and without wear; the notch sides must be parallel with no evidence of where the roller engages; and the guard pin must be straight.

New watch parts are subjected to an Epilame treatment which deposits an invisible coating on selected parts to help prevent oil from spreading. With successive cleanings this coating is likely to be removed, and some repairers re-treat pallets and escape wheels in particular by dipping them into Fixodrop (available through material houses – at a price) and drying with warm air before replacing them into the watch. This is not essential to do, but it is preferred.

27. Replace the pallet.
28. Inspect the pallet cock jewel, and replace the cock and screw. At first, only lightly tack the screw, then check the freedom of the pallet. If free, fully tighten the pallet cock screw and check the pallet is free again. Often the pallet will fall from one banking pin to the other as the watch is rocked. Feel the actual amount of endshake. You should perceive

Fig. 31 Oiling the pallet stones with Moebius 941.

an up and down movement of about 0.03mm on high grade watches, but up to about 5 or 0.08mm on others.

29. Wind the watch by about one turn of the ratchet wheel: this you can do with a screwdriver acting on the ratchet wheel screw. Now check the escapement functions as detailed on p. 44.
30. Bring the exit pallet stone free of the escape wheel and deposit a small spot of special pallet stone oil (Moebius 941) on the impulse face of the stone. (Fig. 31)
31. Allow half of the escape wheel teeth to pass by moving the pallet from one banking pin to the other.
32. Bring the entrance pallet free of the wheel, and put a similar drop of oil on its impulse face. There are alternative ways of oiling pallet stones, but having explored them myself, I feel that the way I have recommended is best.
33. Oil the top pivots of the escape wheel, fourth wheel and third wheel with Moebius 9010, and the top centre wheel pivot with Moebius 9020. Jewelled pallet pivots of watches are not lubricated because it is felt that the liquid friction of the oil is more detrimental than the friction between the pivot and its shoulder, and the jewel hole.

Under inspection, well oiled train pivots should show little evidence of oil in the countersink, but if the wheel is lifted so that the shoulder of the pivot is against the plate, a small fillet of oil should be seen between the shoulder and the jewel. If the bridge were to be removed, there should be a ring of oil around the pivot hole about the diameter of the shoulder of the pivot and with a little height. The pivot itself would have a small fillet of oil in the shoulder. No oil should be present on the top, flat part of the jewel, outside the countersink. (Fig. 32.)

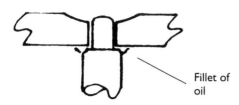

Fig. 32 Classic oiling.

34. If not already done, wash the complete balance in spirit and inspect: (a) the pivots to make sure that they are straight and polished; (b) the balance spring to make sure it is properly formed; (c) the impulse pin to make sure it is clean, upright and secure; (d) the roller to make sure there are no rub marks from the guard pin; (e) that the roller isn't split at its passing hollow or squashed through being driven on forcibly. (Fig. 33)

35. Slacken the stud screw in the balance cock to clear the hole for the stud, and place the cock on the bench with the underside facing up.

36. Lower the balance over the balance cock so that the stud is over the hole and the first coil is between the index pin and boot. Lower the balance, then push the stud further into its hole.

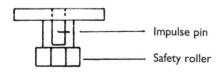

Impulse pin

Safety roller

Fig. 33 Checking the safety roller.

37. Turn the boot to lock the balance spring, and after picking up the cock, tack down the stud screw; this will probably need adjusting later.

38. Turn the cock over gently so that the balance is supported by the spring. Holding the cock so that the balance is over the bottom balance jewel hole, turn the cock so that the impulse pin can enter the notch. Lower the cock to locate the steady pins and withdraw your tweezers. The watch will probably burst into life.

39. Replace the cock screw. While the watch is going, clearly you will not be pinching the balance and so the cock screw may be tightened.

40. Check the endshake of the balance. There must be a minimum of 0.03mm endshake of the balance, but it could be more. The limit has been exceeded if you can feel the end of the pivot catching on the inside of the jewel hole as you move the balance up and down.

Too much or too little endshake can be caused by:

(a) the staff being too short; perhaps due to wear;

(b) the bottom setting being wrongly adjusted;

(c) a bent cock.

The cure for (a) is to change the balance staff. In (b) the bottom setting would be adjusted using the jewelling outfit. In (c) the balance cock is straightened by first removing the balance, then putting it back in the watch, and lowering or raising it with pegwood so as not to mark it.

If the endpieces are screwed, check first that both are secured flat to the plate and cock. It is quite common, especially with learners, to find that endpieces are not properly located and secured. Common faults in older watches include an endpiece that tilts as the endpiece screw is tightened, and a countersunk screw that tightens in the plate but doesn't bite on the countersink in the endpiece.

41. After checking that the distance between the index pin and boot is wider than the thickness of the balance spring, slacken the stud screw again and adjust the height of the stud so that the balance spring is flat. Re-tighten the stud screw.

42. Adjust the balance spring if necessary, so that with the balance in the position of rest, the spring lies half way between the index pin and boot. These should be parallel with only a small gap for the spring to vibrate across – say, one and a half times the spring thickness. The smaller the gap, the better the timekeeping, as the balance amplitude varies. If the watch has the Etachron system, achieve the narrow gap by turning the index pin mount in the regulator.

The amplitude of the balance is the angle the balance turns through measured from the position of rest, to the maximum displacement on one side of the position of rest. The amplitude of the balance diminishes as the mainspring runs down.

43. Check that the watch is in beat. This check could be done on a rate recorder later, or it could be done by hand now. To check by hand, turn the balance with pegwood until drop occurs: as soon as it does, stop turning the balance and release it. The watch should continue to work. Now repeat this with the drop on the other side. (See pp. 42 and 45 for a full explanation of drop and beat setting.)

44. Remove the movement from the holder, turn the watch over and lubricate the bottom scape, fourth and third wheels with Moebius 9010. Leave the bottom pallet pivot hole dry. (Remember jewelled pallet pivots on watches are never oiled.)

45. Oil the centre wheel post with Moebius 9020. (Fig. 34)

46. Replace the cannon pinion. It should snap on and rest against the shoulder of the front pivot.

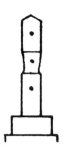

Fig. 34 Oiling the centre-wheel post.

47. Grease the clutch teeth of the winding pinion and the back face, and replace in the watch.

48. Grease the clutch teeth in the clutch wheel and the groove and replace.

49. Grease the small pivot on the stem, the square, groove, main bearing and put a touch of grease on the exposed stem thread, particularly close to the button. Sometimes stems begin to rust there.

50. Feed the stem into the watch, winding pinion and clutch wheel.

51. Secure the stem with the setting lever and screw.

52. Grease the posts for the minute wheel, setting wheel, intermediate setting wheel and yoke.

53. The minute wheel needs to be checked for bent teeth. If a bent tooth is found, simply straighten it with your tweezers. A tooth often get bent down if the cannon pinion is replaced with the minute wheel already in position. Check the setting wheel to see if it has chamfered teeth on one side: if it has, invariably the chamfered side faces the plate. Replace the setting wheel, intermediate setting wheel, minute wheel and yoke. Make sure the yoke passes across the groove in the clutch wheel.

54. Replace the yoke spring after washing in the spirit jar. Place one side of the spring in position, hold the spring down with tissue paper and your finger nail, then load the other side of the spring. After loading, provided the watch hasn't been damaged by rounding corners in the area of the spring, you should be able to remove your fingers and tweezers

without the parts flying across the room. On no account attempt to wind or set the hands otherwise bits will fly.

55. Replace the setting lever spring. Initially just drop it in position, then replace and tack down the screw, load the spring by pushing the end to locate with the stud in the setting lever, and tighten the screw.

56. Lightly grease where the setting lever engages the yoke, where the setting lever spring engages the setting lever, and where the yoke spring engages the yoke.

57. Now try the winding and hand-setting functions.

58. Test that the cannon pinion is sufficiently tight. To do this, pull the stem out, rest the button lightly on your finger, and rock the watch to and fro putting a light clockwise and anticlockwise pressure on the button. Once the play is taken up between the minute wheel and cannon pinion, the button should slide across your finger without the cannon pinion turning at all, even though the minute wheel will rock slightly in both directions. (Fig. 35)

At this point I would rate the watch which is detailed on p. 49. The rate of a watch is the daily change in seconds, so for example, a watch gaining 10 seconds a day dial up would have a rate of +10. In the absence of a rate recorder, if the regulator has not been moved during the overhaul of the watch, leave it where it is, if it is about central; but if the regulator has been moved or removed for cleaning, leave it central for now, which should mean it is no greater than 90° from the stud, and probably no less than 45°. Wind the watch fully, then check the amplitude of the balance. A healthy watch should have an amplitude of up to about 320° when the watch is fully wound, but should drop no lower than an absolute minimum of 220° when the watch has been running for 24 hours. An amplitude of 270° is often quoted as a good amplitude because it lies midway between extremes, and three-quarters of a turn is a convenient number to quote.

If all is well, when the watch is completed, rate it dial up (DU) and button up (PU, PU standing for pendant up) for a pocket watch; and dial up

Fig. 35 Rocking the button to check the cannon pinion tension.

and pendant down (PD) for a wrist watch. These positions are considered the most usual for the respective watches to be worn in. An average gain of about 10 seconds a day between the two positions should be about right.

Never hang a pocket watch for testing in an upright position. The whole watch will swing in sympathy with the balance, and this upsets the rating. Instead, lay the watch at a slight angle to stop it from swinging.

59. Inspect the hour wheel, then place on the corner of a suitable block so that later, when the dial is placed over the hour wheel, the dial feet will clear the block. (The technique we are going to use to replace the hour hand will save unnecessary strain on the movement.)
60. If present, replace the dial washer: it's prime purpose is to keep the hour wheel horizontal, and it is particularly good at preventing the hour hand fouling the small seconds hand, if one is fitted. It would be safe to fit a dial washer anywhere when, after fitting the dial, the endshake of the hour wheel is greater than the thickness of a dial washer.
61. Clean the dial. Mostly this just means a puff with a movement blower, but different dials and different types of dirt mean slightly different treatments. Enamel- and gloss-painted dials will take a gentle wipe with a cloth or gentle rubbing with Rodico, but do not touch luminous paint on luminous dials – quite apart from causing further damage, luminous paint is toxic and must not be swallowed or taken into the bloodstream. Silvered dials are probably best brushed lightly with a fine watch brush. Lacquered dials may be brushed or blown with a movement blower. Raised numerals and battens on dials often improve with a gentle rub over with an impregnated cloth for polishing silver. Be careful not to loosen battens, and to remove any fluff which often gets left behind after cleaning.
62. Place the dial over the hour wheel.

63. Clean the hands. These also often improve with a rub over with an impregnated cloth, particularly gilt hands. If the hands are luminous, the same care is necessary as was recommended for luminous dials.

Re-Luminizing Hands
(Using tritium luminous paint)

(a) Clean the hands up on both sides, taking appropriate precautions with the old luminous material.
(b) Hammer the end of a piece of thin brass pin wire to make a spade end.
(c) After stirring the luminous paint, dip the spade end of the brass wire into the paint, and while holding the hand in tweezers, pass the globule of tritium across the underneath side of the hand; this should fill it. Allow the brass to touch the hand – this will minimize the amount of paint deposited.
(d) Push the tip of the hand into a cork or Rodico until the paint is dry. (Fig. 36)

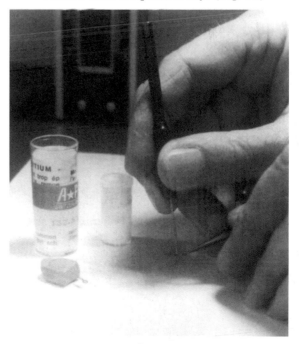

Fig. 36 Re-luminizing watch hands with tritium.

64. Replace the hour hand by placing it on the hour pipe, and then pressing it flush with the top of the hour pipe. This may be done with a special tool, though I often secure the hand with the back of the tweezers. This doesn't mark the hand, and it ensures that the hand is flush with the pipe.

65. Slacken the side dial screw to clear the dial foot holes.

66. Replace the dial ensuring that the teeth of the hour wheel engage the leaves of the minute pinion.

67. Holding the watch in tissue paper to keep fingerprints away, tighten the side dial screws while holding the dial firmly against the movement.

68. Move the hour hand to any convenient hour, and replace the minute hand, pushing it on with a suitable tool made from an old plastic knitting needle. (Fig. 37)

69. When the hand is flat and secure, feel the endshake in the hour wheel: it must have some, and if it doesn't, investigate.

70. Check that the hands are parallel with the dial, clear of each other and clear of the dial.

71. Replace the small seconds hand. Check first that there is no oil on its collet picked up from the fourth wheel, and that there are no hairs or fluff wrapped around the collet.

72. Check that the hand clears the dial all the way around the dial, and that it clears the hour hand.

Replacing the Watch in its Case

1. Make sure the case is clean and free from fluff, particularly if a cloth was used for drying.

2. Remove the stem after slackening the setting lever screw.

3. Put the movement in the case.

4. Insert the button, secure the setting lever screw and observe the freedom of the button

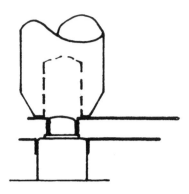

Fig. 37 Pushing hands on with plastic knitting needles.

to turn. This same freedom is necessary after replacing and tightening the case screws. Often the movement turns in the case as the case screws are tightened, causing the button to become stiff to turn.

5. Replace the case screws. Feel the freedom of the button for the reason given above.

6. Replace the case inner and outer backs.

7. Wind the watch fully and set to time.

IN CONCLUSION

Now that you have dismantled, cleaned, checked, reassembled, oiled and adjusted your first typical fifteen-jewel lever pocket watch, you will want to consolidate your new knowledge and skills and build confidence, eventually to tackle smaller and more complicated watches. To this end it is recommended that you work on a number and variety of simple pocket-size watches before scaling down: this means coping with alternative hand-setting arrangements, different winding mechanisms, and different ways of holding the jewels associated with the balance pivots and calendar work. Chapter 4 deals with these points.

3 ESCAPEMENTS AND THE RATE RECORDER

THE CLUB-TOOTH LEVER ESCAPEMENT

The escapement is often spoken of as the 'brain' of the watch, and certainly it is a busy area. However, it is not difficult to understand how it works if what happens is broken down into a sequence of manageable parts. Here we are going to look in detail at the interaction between the impulse pin and notch, the wheel and pallet, and the safety action. (Fig. 38.) To begin with, remember that ideally the balance needs to perform according to the natural laws of springs without outside influence. However, apart from pivoting the balance, another unavoidable outside influence is the impulse to keep the balance vibrating. This impulse is given for part of the time that the impulse pin is in the pallet notch, the remainder being taken up by unlocking the escape wheel. The period when the impulse pin is free of the notch is referred to as the supplementary arc, and that is to be the starting point for our study.

The Impulse Pin and Notch Action

Due to energy within the balance spring, through being coiled or uncoiled, the balance comes out of the supplementary arc, and one side of the impulse pin strikes one side of the notch (Fig. 39). This moves the lever in the direction of travel of the impulse pin, which unlocks the escape wheel, causing the opposite side of the notch to overtake the impulse pin and push it further in the direction in which it is travelling anyway. Very soon the impulse pin leaves the notch, and once more the balance goes into the supplementary arc. In a way that will be seen shortly, after giving impulse, the pallet continues to move but its travel is limited and the notch remains accessible to the impulse pin when it returns.

The amount that the balance turns is limited by the force of the impulse and the strength of the balance spring, but after the balance has turned through something like 270°, the balance spring drives the balance back again and eventually the impulse pin strikes the opposite side of

Fig. 38 The escape wheel, pallet and roller.

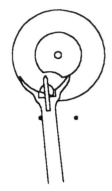

Fig. 39 The impulse pin striking the notch.

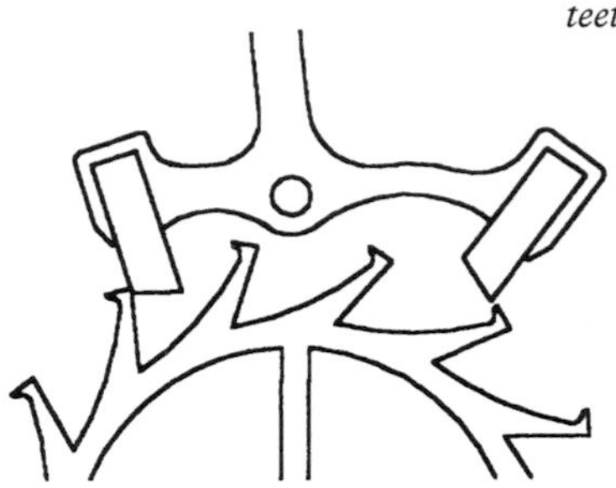

Fig. 40 Pallet embracing 2½ teeth.

the notch; this unlocks the escape wheel to give impulse again and, once more, the (opposite) side of the notch overtakes the impulse pin and gives impulse in the opposite direction. Yet again, the arc through which the balance turns is limited by the strength of the impulse and the force of the balance spring so, after about 270°, the spring drives the balance back again and the whole process is repeated.

The angle through which the impulse pin is in the notch – called the lifting angle of the balance – varies from watch to watch, but is usually approximately 40°. This angle remains constant, whereas the supplementary arc can vary considerably from the watch being fully wound, to needing rewinding after running for twenty-four hours.

The unlocking puts considerable braking on the balance, which explains why, all other things being equal, a watch tends to lose as it runs down. When fully wound, the ratio between the supplementary arc and the combined unlocking and impulse can be about 8:1, but if the amplitude of the balance were to drop to, say, 100°, the ratio would be 2.5:1 – with, of course, the supplementary arc becoming less dominant. Even small changes in supplementary arc can accumulate to several seconds a day change in rate.

Description of the Pallet

The steel pallet frame holds two pallet stones. Integral with the pallet frame is the lever which terminates in the notch. Under the notch is a guard pin, which is part of the safety action to be explained later, and the whole thing pivots about a pallet staff. Most escapements of this type have escape wheels that embrace two and a half teeth, meaning that with the entrance pallet locked, the pallet stones span two and a half teeth of the escape wheel. Each escape-wheel tooth functions with the pallet by first engaging with the entrance pallet stone, and then with the exit pallet stone. (Fig. 40)

Description of the Escape Wheel

In a club-tooth lever escapement, the escape wheel is virtually always made of thin section steel, to help overcome inertia, with the acting face of each escape wheel tooth chamfered to reduce frictional loss between wheel and pallet. The escape wheel very nearly always has fifteen teeth, and is riveted to the escape wheel pinion. Each tooth has a locking face, a locking corner, an impulse face and a discharging corner. (Fig. 41)

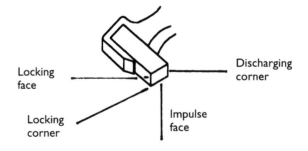

Fig. 41 Faces of a pallet stone.

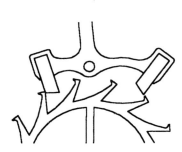

Fig. 42 The escape wheel is locked and the balance is in the supplementary arc.

Fig. 43 Unlocking is complete with a coincidence of both locking corners.

Fig. 44 The various phases of impulse.

Impulse begins

Impulse by the corner of the tooth to the impulse face of the pallet

Coincidence of discharging corner of the stone and the tooth

Impulse given by the face of the tooth to the discharging corner of the stone

Impulse is complete

Wheel and Pallet Action

Our starting point is with the balance in the supplementary arc, the entry pallet stone locking the wheel and the lever resting against a banking pin. (Fig. 42)

Unlocking

The balance comes out of the supplementary arc, and the impulse pin strikes one side of the notch which moves the pallet across so that the locking face of the pallet moves across the locking corner of the tooth. This involves a slight reversal of the escape wheel. Unlocking of the escape wheel is complete when the locking corner of the pallet meets the locking corner of the tooth. (Fig. 43)

Impulse

This part of the action has two phases. The first is where the locking corner of the tooth gives impulse to the impulse face of the pallet, usually occupying five-eighths of the impulse, and the second phase is where the impulse face of the tooth gives impulse to the discharging corner of the stone. This occupies the remaining three-eighths of impulse. (Fig. 44)

At the beginning of impulse, the notch over-takes the impulse pin in the balance, but the full benefit of impulse is not felt immediately due to inertia. It takes time for the escape wheel to start from stationary, and a tooth doesn't really bite into the pallet stone until it is just past the locking corner – which is why wear marks can sometimes be seen on the impulse face at a point just past the locking corner. When the discharging corners of both tooth and pallet stone are adjacent, the impulse is over and the balance enters the supplementary arc.

Drop

With a coincidence of the discharging corners of pallet and tooth, the escape wheel is free to turn through a small angle, nominally 1° to 2½°, and is arrested again by another tooth in the escape wheel dropping onto the locking face of the exit pallet. This is called first lock, the angle of which is measured by taking a line from the pallet centre to the locking corner of the pallet, and a

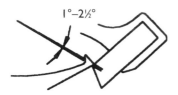

Fig. 45 *First lock with the locking corner of the tooth 1° to 2½° down the locking face of the pallet measured from the pallet centre.*

second line from the centre of the pallet to the locking corner of the tooth. (Fig. 45)

Drop is necessary for the wheel to advance, and takes into account scape and pallet pivot movement in their holes and the inevitable slight inaccuracies of the distance between teeth. Drop is at the expense of impulse. With a fifteen-tooth wheel, as most scape wheels are, teeth are at 24° intervals. As the escape wheel advances half a tooth at a time, action takes place over 12°. As already mentioned, due to the time it takes for the escape wheel to accelerate, the full effect of impulse is not felt immediately and some impulse is lost due to the need for drop. In reality the drop will be the minimum necessary and although on a high grade escapement it could be 1°, it could be as much as 2½° on a lower grade escapement.

When a tooth drops onto the locking face of the entrance pallet it is called 'outside drop' because the face is looking away from the pallet. When a tooth drops onto the locking face of the exit pallet, it is called 'inside drop' because the face lies inside the two pallets.

Run to Banking

Due to the pressure of the locking corner of the tooth against the locking face of the pallet stone – which is inclined by about 15° to a line drawn from the pallet centre and passing through the locking corner on the pallet stone (called the draw angle) – the escape wheel pulls the pallet stone towards the wheel; this involves a slight further turn of the wheel. The angle of draw varies with different escapement designs. (Fig. 46)

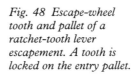

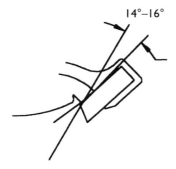

Fig. 46 Angle of draw.

Fig. 48 Escape-wheel tooth and pallet of a ratchet-tooth lever escapement. A tooth is locked on the entry pallet.

The movement of the pallet is limited by the lever striking a banking pin in the plate; in early watches this was a pin in the bottom plate, but in modern horology it is more likely to be limited by the side of the pallet cock which is shaped for the job. This also ensures that the notch is in the right position to receive the impulse pin when the balance comes out of the supplementary arc. The run is over when the lever strikes the banking and the escapement is said to be at full lock. (Fig. 47)

While this was going on, the balance was completing its supplementary arc. Soon the balance returns so that the impulse pin enters the notch once more. Again we have the unlocking, but this time on the exit pallet. There follows the impulse through the exit pallet which again is split between impulse by the locking corner of the tooth to the impulse face of the pallet stone, followed by impulse from the impulse face of the escape wheel tooth to the discharging corner of the pallet, the drop, the first lock on the entry pallet, then run and full lock.

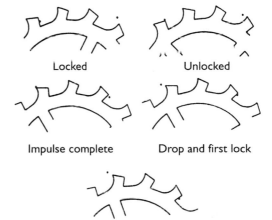

Locked   Unlocked

Impulse complete   Drop and first lock

Run to banking and full lock

Fig. 49 Various stages of action of the pin pallet escapement.

JEWELLED LEVER RATCHET-TOOTH ESCAPEMENT

The action of the ratchet-tooth escapement, found in 'Old English' lever watches, is very similar to the club-tooth lever escapement, the significant difference being that only the pallet has an impulse face: the tooth simply has a point of about ½° thickness to give it strength; it is made of brass. Apart from this, the action is much the same as with the club-tooth lever escapement, with unlocking, impulse, drop, first lock, and run and full lock. (Fig. 48)

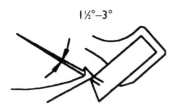

Fig. 47 Full lock.

PIN PALLET ESCAPEMENT

In the pin pallet escapement – found in lower-priced watches – it is the escape wheel that has the locking and impulse faces: these act on pallet pins – usually round pins raised into the vertical plane. The action is almost identical to the jewelled lever club-tooth escapement. Run to the banking can be limited by banking pins, or by a pallet pin contacting the root of a scape-wheel tooth. The various stages of the action are shown in Fig. 49.

SAFETY ACTION

Being worn about the person, a watch is likely to get knocked many times during the day, with the concomitant risk of accidental unlocking of the escapement, and the pallets passing to the other side of the balance, leaving the notch in the wrong position to receive the impulse pin when the balance comes out of the supplementary arc. If such a condition does occur, the watch is said to be overbanked. To prevent overbanking, in the case of the jewelled lever club-tooth escapement (and often the pin pallet variety of club-tooth escapement), a guard pin is fitted close to, and usually under the notch, to act in conjunction with a safety roller on the balance staff. (Fig. 50)

An Old English lever usually has a large single roller which carries the impulse pin and has a

passing hollow. The guard pin extends from the top of the pallets near the notch. This arrangement has the disadvantage of a greater surface area and therefore the potential for greater frictional losses; also it is easier for the roller to drag the guard pin, and with it the pallet, to the wrong side causing overbanking.

With each of the above, in the event of a shock, the guard pin touches the safety roller preventing the pallets from crossing to the wrong side. As an escape-wheel tooth is still pressing against the locking face of a pallet stone, run to the banking occurs a second time taking the guard pin away from the safety roller in the process. With this arrangement, it is unlikely that even one beat will be lost.

CHECKING THE ESCAPEMENT

After reassembling a cleaned watch, before the balance is replaced, carry out the following checks:

1. Put a little power on the watch, winding by just one or two turns of the ratchet wheel. Lead the lever from one banking pin to the other, observing that the first point of contact between tooth and pallet stone after drop is on the locking face and safely below the locking corner. About 1½° would be considered safe. This is purely a visual test, that the locking looks safe and is not too deep or too shallow.

2. Lift the lever away from the banking pin, but not so that the tooth comes onto the impulse face of the stone, then release the lever and observe that the run to banking occurs. Do this on both entry and exit pallet stones.

3. Push the lever off the banking pin sufficiently for impulse to occur, and observe the pallet jump smartly to the other banking pin. Repeat this from the other banking pin. Don't be concerned if the pallet hits the opposite banking pin and bounces back to the same banking pin: it is merely

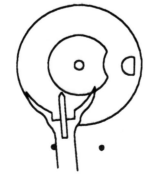

Fig. 50 Guard pin, safety roller, horn and notch.

rebounding because it is moving more quickly than it will when giving impulse. It is therefore unlikely to be a problem in the going watch.

4. After replacing the balance, check the 'shake on the lever'. This means holding the balance stationary after the passing hollow in the safety roller has passed the guard pin, and pushing the lever across so that the guard pin touches the safety roller. There must be movement of the lever and it should be about half a degree, during which time the scape-wheel tooth must remain on the locking face of the pallet stone. The lever must return to the same banking pin when released. The same test must be carried out when the lever and notch is on the other side of the safety roller.

5. Check that there is minimal play of the impulse pin in the notch.

6. Check that as the balance comes out of the supplementary arc, there is clearance between the impulse pin and the horns on each side of the notch.

CHECKING THAT THE WATCH IS IN BEAT

Definition
A watch is said to be 'in beat' when the movement of the impulse pin is the same on each side of an imaginary line joining the balance and pallet centres. Alternatively you could say that when the balance is in the position of rest, the impulse pin should lie on the imaginary line joining the balance and pallet centres. If it does not, the watch is said to be 'out of beat'. The significance of being 'in' or 'out' of beat is that if it is in beat, the watch is much more likely to self-start on winding, and any chance of the impulse pin striking the wrong side of the notch as the watch works normally – a condition known as 'knocking' – is minimized. Being in or out of beat is about the relative position of the collet and impulse pin.

Testing for Being 'In Beat'
1. With a partially wound watch, turn the balance under control until drop occurs: for clarity let us assume that this happens on the exit pallet stone.

2. Stop turning the balance immediately after drop, so there is no supplementary arc, then release the balance. It should turn in the opposite direction to the way you were turning, and the watch should continue to work.

3. Do exactly the same again, this time allowing drop to occur on the entry pallet stone, again without supplementary arc.

4. Release the balance, and again the watch should continue to function.

If the watch 'goes' from both sides, or in fact neither side, the watch is sufficiently in beat to function properly. If, however, after releasing the balance it doesn't turn in the required direction from just one side, the watch is out of beat and will need putting in beat.

Putting in Beat
This means correcting the relative positions of the impulse pin and collet. Generally it involves holding the balance, and therefore the roller and impulse pin, stationary while turning the collet in the opposite direction to that in which the balance should have turned. (Or if you prefer it, hold the collet stationary and turn the rest of the balance assembly in the direction in which the balance should have turned upon release.) Turn by a little at a time, say 5° or less, until the balance functions normally from whichever side the test is made. For adjusting the beat, I use a balance cock stand and a collet removing tool. Take extra care not to damage the inner coil of the balance spring as the tool is entered into the split in the collet. (Figs. 51 and 52)

Some watches have a moveable stud holder; if that is the case, simply turn it in the direction in which the balance should have turned on release.

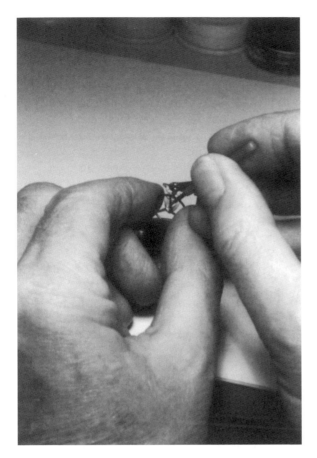

Fig. 51 Putting in beat.

THE RATE RECORDER

A rate recorder, or 'timing machine' as it is often called, is not essential for watch repairing, but it certainly does speed up the process of checking the likely performance of a watch, saving several days of testing in a number of positions. It is useful for regulating a watch and for fault diagnosis. More correctly the rate recorder should be called an 'instantaneous rate recorder' because it tells you what the watch is doing right now, rather than what will happen over the next twenty-four hours.

For a competent repairer, the greatest use made of the rate recorder is simply to rate the watch, and in my experience little use is made of it until after a watch has been overhauled when virtually all faults have been found and attended to, leaving only the rating to do. However, on occasions it is most useful in helping to track down the more obscure faults that sometimes present themselves after the watch has been overhauled.

The rate recorder can be thought of as a master watch against which you will compare the performance of the watch you have been working on. It has an acoustic piezo-electric 'pickup' which detects and amplifies the sounds made in the watch, compares them with its own oscillator, and feeds out intelligence in the form of a printout which you will be able to interpret. (We will be looking at the Greiner Micromat, but other rate recorders for mechanical watches are similar.)

In order to do this, paper from a roll – a bit like a till roll – is fed between rollers at a fixed speed. A synchronous (meaning 'in step with') motor drives what looks like a drum with a single screw thread spiralling over its surface. The speed at which the drum rotates is selected according to the number of vibrations per hour of the watch balance. Above this is a recording bar with ink, and with each beat of the watch, the bar drops momentarily, trapping the paper between the recording screw and the recording bar.

If the signal arrives on time, a straight line is printed on the paper roll. If it arrives a little early with each tick, the printed line leans to the right; and if it arrives a little late, the line leans to the left. The amount of lean determines by how much the watch is losing or gaining, and can be read off by aligning grid lines on a plastic disk with a scale printed on the top housing. (Fig. 53)

Fig. 52 Beat-setting and collet-removing tool.

Fig. 53 Grid lines on a Greiner Micromat rate recorder.

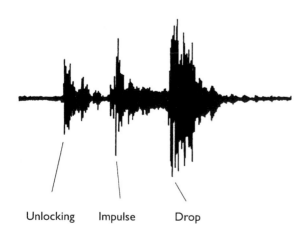

Unlocking   Impulse   Drop

Fig. 54 Unlocking, impulse and drop seen on an oscilloscope.

There are three main signals that can be picked up by the rate recorder; these are: (a) the unlocking signal that is largely the impulse pin striking the notch, but includes the locking face of a pallet stone moving across a tooth; (b) the impulse, made up of the sound of the impulse pin in the notch and the action between a tooth and a pallet stone; and (c) the drop signal, which is a tooth dropping onto the locking face of a pallet and the lever striking a banking pin. Fig. 54 shows what these three signals would look like on an oscilloscope.

For all rate measurement the volume control is adjusted for the unlocking signal: being the weakest signal of the three, the volume control needs to be turned about half way up. Although the other signals are present, the rate recorder will ignore them. By turning the volume control back, the impulse signal can be picked up, and a further reduction in volume enables the drop signal to be picked up. As the volume control is turned up and down, often all three signals are clearly identifiable. The print-out will be stepped because each signal arrives at a different point in time. (Fig. 55)

Clearly there are an infinite number of positions a watch could be rated in. For practical purposes the full range of positions is dial up (DU), dial down (DD), pendant right (PR), pendant down (PD), pendant left (PL) and pendant up (PU). For ordinary commercial work, rating is often carried out in just two positions, namely those considered to be the positions in which the watch is most frequently worn. For a pocket watch these positions would be DU and PU; for a wrist watch, DU and PD. (In practice, watches are often first brought to time DD, as this leaves the regulator accessible. However, it is necessary to check that DD and DU are very similar before relying on this measurement.)

For ordinary commercial work, the objective with a reasonably good quality watch is to get a gain of about 10 seconds a day as an average of two positions. Final regulation may have to be achieved after the customer has been wearing the watch. The eventual performance of a watch

	Volume control position	Oscillogram	Trace

Fig. 55 Stepped trace as the volume control is reduced.

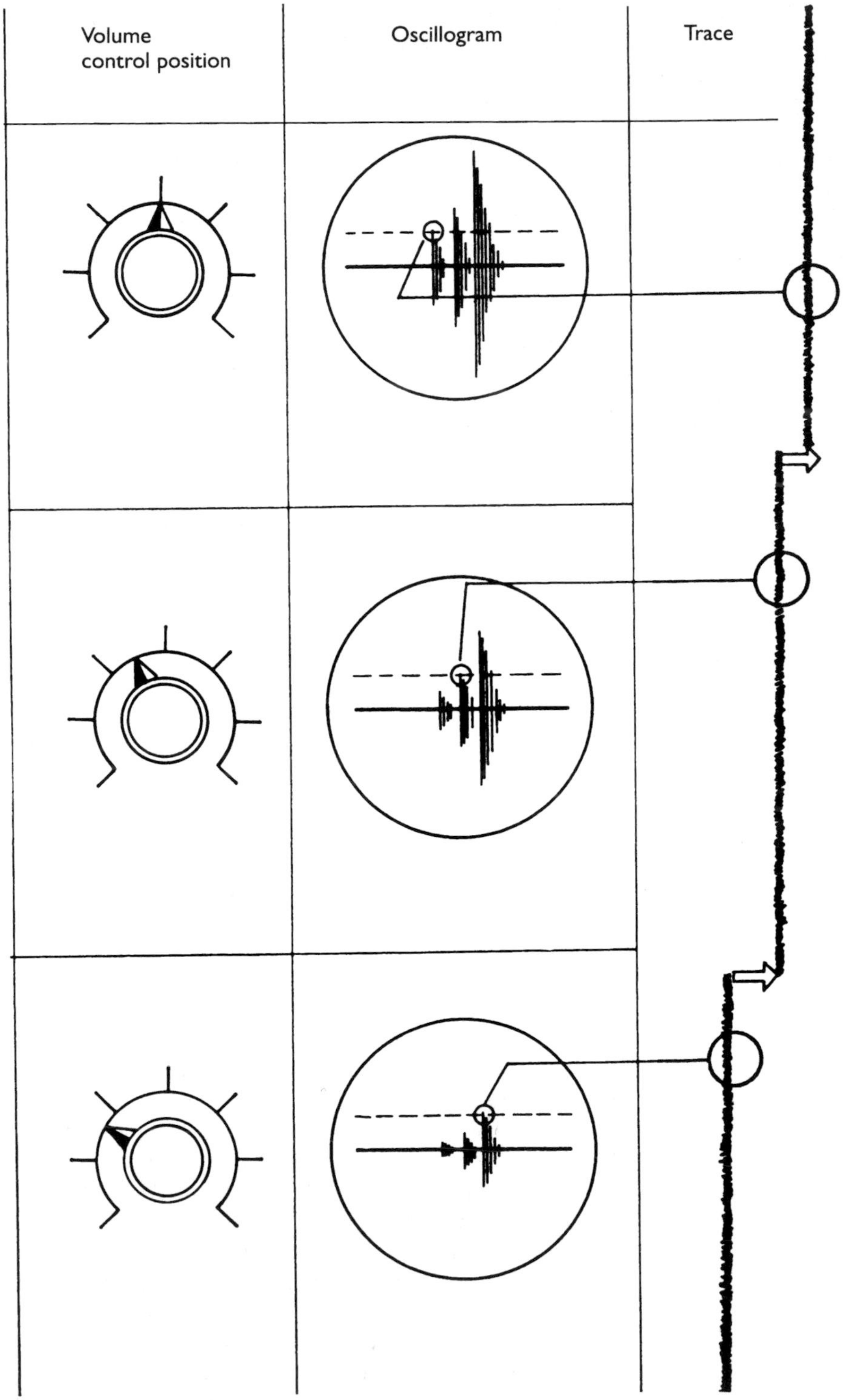

is limited by design and materials, but it is also influenced by its environment. An ordinary watch is likely to have a different average rate with different people and different seasons. It is quite common for watches to gain in winter and lose in summer due to changes in elasticity (springiness) of the balance spring.

RATING A WATCH

1. To prepare the watch for rating, wind the watch fully, then let the mainspring down by about a turn of the ratchet wheel. Here it is assumed that the watch has been overhauled, properly inspected and repaired as necessary, oiled, and preliminary adjustments made.
2. Position the watch on the pickup DD.
3. Turn the rate recorder on, select the frequency at which the balance vibrates per hour, adjust the volume control to about half to pick up the unlocking signal, and position the watch on the pickup. After establishing the rate, write down the figure as + (plus) or – (minus), then turn the rate recorder off or lower the volume control just to save paper. Let us assume that DU the rate is +16.
4. Turn the watch to one of the vertical positions, PU for a pocket watch, PD for a wrist watch, and wait about 20 seconds for the amplitude of the balance to settle.
5. When it has, turn on the rate recorder and write down the rate in this second position. Here we will assume it is a pocket watch and the PU rate is –4.

The average rate between the two positions is (+16–4)/2=6 so the watch is expected to gain about 6 seconds a day.

Should you be rating the watch in more positions, add the plusses together, add the minuses together, then take the minuses from the plusses and divide by the number of positions you used for rating.

In a healthy watch, the frictional losses DU and DD are expected to be less than the frictional losses in the average of PR, PD, PL and PU positions. This is due to the change in friction when the weight of the balance is on the end of the pivot, and when it is on the sides of the pivot. This is reflected by a change in amplitude of the balance, so expect the amplitude to be higher DU and DD than it is with the watch in the vertical positions. (Fig. 56)

When the watch is DU or DD (we never say horizontal) we say that the watch is in the 'long arcs'; when it is in the verticals (trade language for PR, PD, PL and PU) it is said to be in the 'short arcs'. The reason for the comparative loss in the short arcs as compared with the long arcs is that when it is in the short arcs, due to the slightly lower amplitude of the balance and thus the supplementary arc, the braking effect of the

Fig. 56 Shows why the frictional losses in the vertical positions are greater than those in the DU and DD.

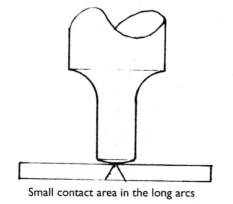

Small contact area in the long arcs

Large contact area in the short arcs

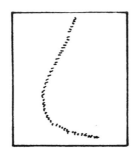

Fig. 57 Printout showing the rate as the amplitude picks up.

unlocking action causes a comparative loss. To summarize, expect the average of the short arcs to be slow when compared with the long arcs.

The above can be demonstrated by putting a wound watch on the rate recorder DD, stopping the balance then starting it again, and observing the change in rate as the balance amplitude picks up. Typically the printout will be as shown in Fig. 57.

Out-of-Poise Effect

Being 'out of poise' means that the balance, which includes the roller, collet and balance spring, has a heavy spot. When the watch is in the long arcs (DU or DD), any out-of-poise of the balance does not register. In the short arcs, out-of-poise shows as significantly different rates in different vertical positions. Static poising should bring differences due to poise errors closer than 30 seconds a day. (*See The Clock Repairer's Manual* for static poising.)

Rating an Actual Watch

A typical fifteen-jewel lever pocket watch has been overhauled (the one used in Chapter 2); it has been wound fully, then let down by one turn of the ratchet wheel. It was put on the rate recorder with the following outcomes:

DU +40	PR +40
DD +35	PD +40
PL +32	
PU +30	

Thus long arcs average +37.5 sec/day; and short arcs average +35.5 sec/day.

These results are very good indeed for an ordinary commercial watch. We have just a 5-second difference between DU and DD; 10 seconds a day as the greatest difference between vertical positions; and only two seconds difference between the averages of the long and the short arcs.

In practice we would do nothing further to this watch except regulate to have a gain of about 10 sec/day. However, if we chose to, we could close the gap between DU and DD and have two options. In the DU position, the balance is resting on the end of the top pivot. In the DD position the balance is resting on the bottom pivot. Our choices then are (a) to make the balance top pivot blunter, thereby increasing the frictional losses and bringing the DU rate closer to the DD; or (b) to sharpen the bottom pivot to decrease the DD friction, thus bringing the DD rate closer to the DU. (Complicated, isn't it?)

Because the average of the long arcs is fast on the average of the short arcs, (a) is the better option because not only does it bring DU closer to DD, but it also brings the long and short arcs closer.

DYNAMIC POISING

When a balance – which includes the roller, collet and balance spring – has a heavy spot, we say that the balance is out of poise and we assume that the heavy spot is in the rim of the balance. Effectively it is. In the long arcs, this out-of-poise is not evident, but in the short arcs, the rate differs in different positions.

In theory at least, if we put a wound watch on the rate recorder in a vertical position, we could rotate the 'pickup' and discover the position in which the watch had the greatest loss. Assuming there were no other faults with the watch, if we were now to stop it so that the balance were in the position of rest, we would have the heavy spot at the bottom of the balance.

Once again in theory, we could lighten the heavy spot, or make the top of the balance heavier, to put the watch in poise. In practice,

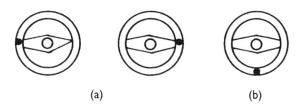

(a)   (b)

Fig. 58(a) With the balance in the position of rest and the watch in a vertical position, any out-of-poise at the 9 o'clock or 3 o'clock position is self-cancelling. (b) With the balance in the position of rest and the watch in a vertical position, the watch will show a loss compared with other vertical positions at amplitudes over 220°.

once the heavy spot is confirmed, we do one or the other or both, though it is more usual just to take the rate in four vertical positions rather than find the actual point with the greatest loss. With the pendant up and the balance in the position of rest, the screw at the bottom of the balance is called 'the pendant up, down screw'. The naming of the other three vertical positions is similar, so there is a pendant right, down screw; a pendant down, down screw and a pendant left, down screw.

If we take any one vertical position just for the moment with the balance in the position of rest, a heavy spot at the bottom will cause a loss in that position, a heavy spot at the top of the balance will cause a gain, and a heavy spot to the left or right has no effect on the rate. (Fig. 58)

The reason for the loss in the vertical position when the heavy spot is at the bottom of the balance is that at impulse, the heavy spot has to be driven against the pull of gravity in addition to the effect of unlocking the balance. As you might imagine, a heavy spot at the top of the balance would lead to a gain, because gravity acts on the heavy spot, effectively contributing to the impulse.

What is also interesting is that at an amplitude of 220° the effect of out-of-poise is nullified, but at arcs below 220°, the effect is reversed so that a loss becomes a gain but the gain is exaggerated. We can use this to prove that errors are due to

out-of-poise by rating the watch at a good amplitude of 270° and again at an amplitude of 180°. Rating the watch at the higher amplitude is known as the 'up' rates; at the lower amplitude it is known as the 'down' rates. With a heavy spot at the bottom, a small comparative loss at 270° should become an exaggerated gain at 180°. (Fig. 59)

Once the heavy spot is confirmed, in a lower grade watch (but still good), if the balance has screws, the head of the screw could be reduced with an escapement file or a timing washer could be put underneath the head of the opposite screw, or both.

When you lighten a screw, all the rates increase, but the greatest increase will occur in the position where you had a heavy spot. Whether a balance is lightened or made heavier will depend on which will leave the regulator nearer to the central position on the balance cock.

If the rim of the balance is smooth, a little metal is removed from underneath the balance at the heavy point with a small watch drill. It is not practical to make a screwless balance heavier.

Dynamic poising is really better left to the experienced repairer, and even then it is only likely to be done on high grade watches where perhaps adjustable timing screws are provided. For ordinary watches, static poising is likely to be quite sufficient. No attempt should be made to dynamically poise a watch that is not capable of responding to the treatment.

ISOCHRONAL ERRORS

It is interesting to see the printout of a watch as the balance amplitude varies: this shows that at different amplitudes, the time of vibration of the balance varies. If differing arcs were all performed in the same length of time, we could say that the balance was isochronous.(Iso means equal, chronous means time.) Isochronism is unlikely to be achieved, but in modern watches with a flat balance spring, the design of the

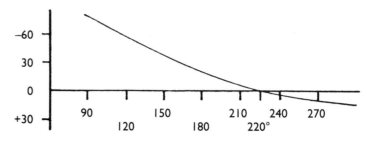

Fig. 59 Graph showing
how an out-of-poise
balance affects the rate.

watch is conducive to isochronism and the repairer is encouraged to leave only a small gap between the impulse pin and boot to make the balance as isochronous as it can be.

Making this gap narrow ensures that for a greater part of the time, only that part of the balance spring from the collet to the regulator affects the rate. As the mainspring runs down, the balance spring spends more time touching neither index pin nor boot, so that increasingly the part of the balance spring between the regulator and stud influences the rate, thus upsetting the isochronism. The part of the balance spring between the regulator and stud has virtually no effect on rate while the balance spring rests against the index pin or boot.

Typical 'Printout' Interpretations
Below are typical printouts and their interpretations. The apparent single line in the first printout is in fact two lines superimposed: one line relates to the entry pallet, the other the exit. These will show up as two separate lines when the watch is out of beat and the distance between the lines will vary with changing and poor amplitudes.

Two separate lines don't always mean the watch is out of beat. I have learned to put faith in the hand test above reliance on two separate lines as indicators of being in or out of beat (Fig. 60).

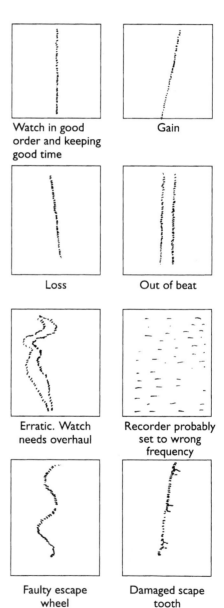

Fig. 60 Interpretations
of different traces.

ALTERNATIVE PROCEDURES

Alternatives for Hand-Setting, Winding, Supporting Balance Pivots, Calendar Work

ALTERNATIVES FOR SETTING THE HANDS

Already we have seen the typical centre wheel with backslope and snap-on cannon pinion. In some early watches the centre wheel is hollow, and a squared hand-set arbor is passed through the centre wheel and carries a cannon pinion. The cannon pinion is a friction drive onto the arbor, leaving no endshake between the hand-set arbor and the centre wheel – but of course the centre wheel itself does have endshake between the plates.

To remove the cannon pinion, tap the end of the hand-set arbor smartly with a light watch hammer while holding the movement in your thumb and finger tips. Once the end of the arbor is flush with the top of the cannon pinion, this should lift off and the hand-set arbor should pull out.

To replace it after the train has been assembled, oil with 9020 the section of the arbor that contacts the inside of the centre wheel, and push the arbor into the centre wheel all the way, checking that the friction between the two is sufficient for the hands to carry. If it is not, increase the friction by resting the arbor on a block and swelling a small area that engages with the inside of the centre wheel with a small flat punch and watch hammer.

Next, the hand-setting square is rested on a stake, and the cannon pinion is driven on with a flat hollow punch, not to increase the friction, but just to take up any endshake between arbor and centre wheel. The final set-up is shown in Fig. 61.

In typical 13''' (29.25mm) Roskopf movements, which were common in the 1950s and

earlier, the cannon pinion did not snap on, but instead was placed onto a post on the front of the bottom plate. The watch had no centre wheel: instead the cannon pinion was driven by what was in fact the minute wheel, which was lightly riveted to the barrel or barrel cover. The minute pinion was fixed to the minute wheel for driving the hour wheel. (Fig. 62)

The barrel, or cover, had a brass boss riveted securely to it, but with a shoulder to accommodate the minute wheel and pinion, and with a hole to accept the barrel arbor. The minute wheel and pinion were placed over the shoulder, then lightly riveted with a round-ended punch to spread the rivet to form a friction drive. (Fig. 63)

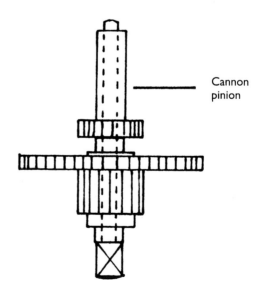

Cannon pinion

Fig. 61 The canon pinion arrangement on an early pocket watch.

Fig. 62 Roskopf movement where the minute wheel drives the cannon pinion which 'drops' over a post in the plate.

After cleaning, the clutch arrangement must be lubricated at the points shown, and then tested. A slack friction drive is easily tightened, but do it in modest stages. An over-tight friction drive is not easy to loosen.

Contemporary Cannon Pinions

Very often, modern cannon pinions still drop over a post, there being no conventional centre wheel, and the clutch arrangement is between a wheel with three legs in the centre and a steel cannon pinion; alternatively it may be fitted to the barrel. Three variants are shown in Fig. 64. Sometimes the friction arrangement is to be found on a train wheel (the second wheel in fact).

When overhauling different watches it is essential to locate the clutch arrangement which allows hand adjustment to be made. Find it, lubricate it with pressure-resistant oil or grease, and test it. Sometimes a cannon pinion can be tightened; at other times a new cannon pinion has to be obtained.

Alternatives to Shock-Proofing Balance Pivots

When overhauling a variety of older pocket watches, you will encounter two main arrangements for holding jewel holes and cap jewels. The first is where the jewel holes are secured directly in the balance cock and the bottom plate, with the cap jewels (endstones) secured to endpieces; and the second is where the jewel holes and cap jewels are mounted in what looks a bit like bushes – these are called mounted jewels. (Fig. 65)

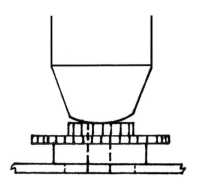

Fig. 63 *The minute wheel and pinion being riveted to the barrel to form a clutch arrangement for a friction drive.*

Mounted jewels must be pushed out and in by pushing with pegwood on the brass mount, and not on the jewel itself. Orientation of the mounts for the cap jewels is ensured by a small pop mark or countersink which can be seen from the top of the cock, overlapping the cock and the mount. The reason for this is that the endpiece mounts are secured by two screws which are not necessarily exactly opposite one another.

Often the screws for mounted jewels have stripped threads, and problems with these can often be overcome in the following way: support under the head of a screw after it has been replaced, and tap the end of the threaded part of

Fig. 64 *Three contemporary cannon pinions.*

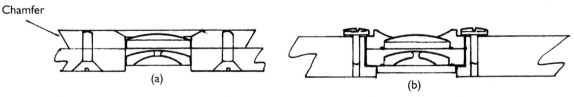

Fig. 65(a) *Top endpiece with a chamfer to accept the regulator and two screw holes. The bottom endpiece has only one screw hole and no chamfer. (b) A mounted jewel.*

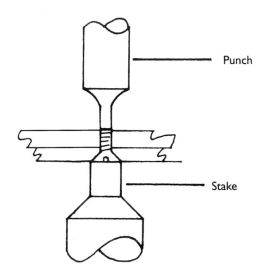

Fig. 66 Supporting an endpiece screw, while the thread is expanded to grip the brass.

the screw with a pin punch as if riveting. This spreads the screw a little so that the good thread in the screw expands slightly to tighten on the brass of the cock or bottom plate. Avoid riveting tightly because you will probably have to dismantle the whole thing again one day. (Fig. 66)

When reassembling the mount in the plate or balance cock you may find that they are a little tight. Assist them into position by pushing on the brass with pegwood. Correct positioning should avoid problems with balance endshake.

WINDING MECHANISMS

On early watches, winding and hand-setting was achieved by a key which was quite separate from the watch. On many of these watches, a pusher in the pendant had to be depressed, which caused the hinged and sprung back to open. An inner back was then revealed with two holes, one for the winding square, the other a square for hand-setting (this type of hand-setting has already been described: see p. 53).

An alternative was for both front and back of the watch to be hinged. Opening the back gave access to the winding square, whilst the hinged bezel gave access to a square at the top of the cannon pinion for hand-setting. The minute hand was fitted to the same square.

The quest was on for winding and hand-setting which did not require a separate key, and various devices were conceived which have the generic term 'keyless work'. The arrangement described in the overhaul of a typical fifteen-jewel lever watch is a type of keyless work known as 'positive setting'. Others include 'push piece', 'rocking bar' and 'negative setting'. A common variety of each is described.

Push Piece
This is very similar to positive setting, with a clutch wheel, winding pinion, yoke, yoke spring and a stem and button – except that instead of pulling the stem out for hand-setting, a pusher would be operated by the thumb nail which pushed directly on the yoke. (Fig. 67)

Rocking Bar
With this type of winding and hand-setting, there is no ratchet wheel or crown wheel to be seen from the back of the watch; instead there is a 'ratchet wheel lower' between the barrel and front plate. There is no crown wheel, nor a clutch wheel. Instead, the winding pinion has a hole which is square – or at least keyed in some way – so that it always follows the winding stem irrespective of the direction in which it is turned.

The winding pinion, when turned, turns three transmission wheels held under a rocking bar. One of these wheels engages with the 'ratchet wheel lower' for winding. When the stem is pulled out, the setting lever causes the rocking bar to rock or rotate slightly, disengaging one of the transmission wheels with the 'ratchet wheel lower', but putting a different transmission wheel into engagement with the minute wheel for hand-setting. When the stem is pushed in, a return spring pushes on a post under the rocking bar to return it to the wind position again. (Fig. 68)

Fig. 67 Push piece.

*Fig. 68
Rocking bar.*

When reassembling the rocking bar, grease the three posts over which the three transmission wheels fit: this will not only lubricate the centres of the transmission wheels, but it will also hold the wheels in position on the rocking bar while it is turned over and fitted to the watch. Also grease the pin for the return spring, and the pivoting point of the rocking bar.

Negative Setting

There are many variants of negative setting, but the one thing they all have in common is that when the movement is removed from the case, the watch automatically goes into the hand-setting mode. The variety to be described here has a clutch wheel and winding pinion, a two-part stem, a shipper which engages with the clutch wheel and does a similar job to a yoke, and a shipper lever which engages with the stem and acts a bit like a setting lever. The whole arrangement is held down with a cover plate. (Fig. 69)

With the movement removed from the case, the shipper spring causes the clutch wheel and winding pinion to separate by the interaction of the shipper and shipper lever. This is shown with the cover plate removed. (Fig. 70)

By pushing the part stem in, the clutch wheel and winding pinion come together for winding the watch in the usual way. Again, this is shown with the cover plate removed, in Fig. 71.

The 'part stem' is pushed in by the other half of the stem which has the button attached to it (Fig. 72). To function correctly, the part of the stem which carries the button has to be held in two quite positive positions. This is done by a sleeve screwed into the pendant of the watch. (Fig. 72a)

Fig. 69 Negative setting with cover.

Fig. 70 *Clutch wheel and winding pinion separated. Cover plate removed.*

Fig. 71 *Winding with cover plate removed.*

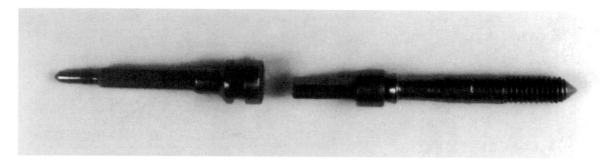

Fig. 72 The two-part stem.

Removing the Sleeve

This is self-evident from fitting and adjusting the stem. Brief instructions for removing the sleeve are:

1. After removing the movement from the case, hold the square of the part stem with brass pliers and unscrew the button.
2. Select the appropriate key from the sleeve key, insert it into the pendant and sleeve, and unscrew the sleeve.
3. Separate the part stem from the sleeve.

Fitting and Adjusting the Sleeve

1. By hand, screw the sleeve into the pendant, making sure that the thread is not crossed.

2. When the top of the sleeve is flush with the top of the pendant, give it about another two turns with the sleeve tool; the thread should begin to tighten up slightly. If the thread of the sleeve is too loose in the pendant, support its top from underneath and then spread it with a rounded punch.
3. Replace the stem from the inside of the case.
4. Feed the movement, with the dial off, into the case so that one part of the stem enters the other part. Secure the movement with its case screws. If this is not done, the movement may occupy a slightly different position in the case, resulting in an incorrect adjustment.
5. Operate the stem two or three times to make sure everything is located properly. At this

Fig. 72a Stem with the sleeve.

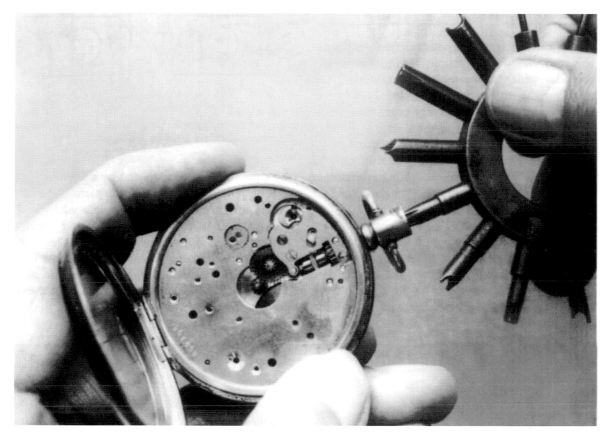

Fig. 73 Screwing the sleeve in with the sleeve tool.

stage, when the stem is pushed in and released, the clutch wheel and winding pinion should not fully engage.

6. Now adjust the sleeve by screwing it in with the sleeve tool until there is full engagement between clutch wheel and winding pinion. Test as the sleeve is lowered. Do not screw the sleeve in by more than is necessary. (Fig. 73.)

7. The adjustment is now made, so remove the movement from the case, hold the winding square with brass-lined pliers, and screw the button onto the stem and replace the dial and so on.

Sometimes a sleeve gets broken and will need to be replaced. The symptom of a worn or broken sleeve is that when the stem is pushed in after hand-setting, it is pushed out again by the shipper spring.

Some negative setting watches are provided with a means of locking the clutch wheel and winding pinion together for bench testing out of the case. In the example shown, the two-part stem is locked down by a screw with a passing hollow. In the normal mode, the passing hollow is adjacent to the part stem. When out of the case, the part stem can be locked down by turning the screw. (Fig. 74)

There are other systems, including a sliding bar which pulls the clutch wheel up into engagement with the winding pinion; this system is self-explanatory.

CALENDAR WORK

Many watches are fitted with calendar work, and this may vary from a simple date to a perpetual calendar which automatically takes account of twenty-eight, thirty and thirty-one day months and even leap years, occurring every fourth year when February has twenty-nine days. In simple form, date changes may be gradual, taking something like 1½ hours to change; or they may be instantaneous, going from one date to the next at midnight.

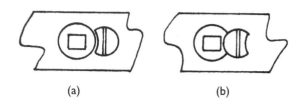

Fig. 74 *Locking down the part stem. (a) Normal mode. (b) Part-stem held down for bench-testing.*

Gradual Date Change

Figs. 75 and 76 show a simple gradual date-change mechanism, as found in many watches.

When the dial is in position, the correct date can be seen in the window at the three o'clock position. The date indicated is held in position by a pivoted jumper assisted by a jumper spring positioned in a recess in the plate. The hour wheel drives the intermediate date wheel which has two sets of teeth: while the upper teeth are driven by the hour wheel, the lower teeth drive the calendar driving wheel. A raised post in the calendar driving wheel engages with one of the thirty-one teeth in the date indicator once every twenty-four hours to change the date.

Fig. 75 *Gradual date change with guard.*

Fig. 76 Gradual date change without guard.

Initially the date indicator is turned by the raised post in the driving wheel. Eventually the trailing edge of a tooth in the date indicator advances beyond the point of the jumper, whereupon the jumper pushes the date indicator and locks between two teeth again. This reduces the length of time that the date takes to change over.

Dismantling
To dismantle this type of arrangement, loosen the screws holding the date indicator guards, lift the date indicator close to the jumper and allow the jumper to go under the date indicator. This releases most of the tension in the jumper spring, reducing the chance of it 'flying' out and getting lost. With the tension off the spring, continue to unscrew the screws holding the date indicator guards, and carefully remove them. Hold one side of the jumper spring with tweezers and care-

fully remove it; I usually rest a fingernail lightly on the crook of the spring to control it.

Reassembly
After cleaning, lightly grease the outside of the cannon pinion where the hour wheel bears, the post of the intermediate date wheel, the post of the calendar driving wheel, and the pivot of the jumper. Replace the calendar components, allowing the jumper to rest under the date indicator. Replace the jumper spring using fingernail and tweezers again, then carefully replace the date indicator, the date indicator guards and screws. Lightly tack down the screws, pull the jumper back so that the date indicator drops fully, then release the jumper so that it rests between teeth, and begin tightening the retaining screws. Tighten them fully only when you are sure nothing is trapped.

Grease the tip of the jumper, three or four teeth of the date indicator, the raised post in the calendar driving wheel, and where the jumper spring touches the jumper. Replace the hour wheel, and try the mechanism by turning the hands – making sure that the date does change properly.

After replacing the dial, replace the hands so that the date begins to change at 11.30 and completes the change at some time after midnight. To correct the date it is usually unnecessary to turn the hands through twenty-four hours: simply allow the date to change, turn the hands back about four hours, then forward again. I usually replace the dial with the date retarded by about four days to the actual date, to minimize the number of days that the date has to be advanced to show the correct date. This allows for replacing the hands in the right place, and for checking.

Many watches with calendar work have a facility for rapid advancement of the date and also the day, where this is fitted. The stem has three positions: one is when button and stem are fully pushed in, and wind the watch as usual. Two describes the first pull out of the button, when the watch continues to function normally but the date and day may be advanced rapidly. Three is the second pull out, which allows normal hand-setting, but the calendar advances once every twenty-four hours at about midnight (the same as would happen without rapid advancement of the day/date).

Instantaneous Date Change
In these watches the date changes instantly at midnight (Fig. 77). In this calibre, the hour wheel has a double set of teeth, the lower set driven by the minute wheel pinion, the upper set driving the calendar driving wheel. The

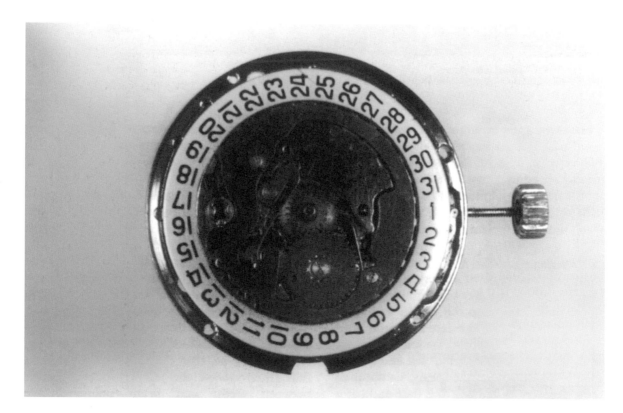

Fig. 77 Instantaneous date change mechanism.

Fig. 78 Underside of the calendar driving wheel showing the cam.

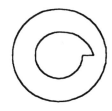

underneath side of the calendar driving wheel has a cam fixed to it, as shown in Fig. 78. As this wheel turns, the cam lifts the unlocking yoke for the date indicator, which carries a spring-loaded, pivoted pawl which displaces as the unlocking yoke lifts: this displacement prevents the date indicator turning backwards. When the pawl passes the tooth, it springs back into place ready to drive the date indicator around by one tooth. The unlocking yoke continues to be lifted until the dropping-off point of the cam is reached. Once the unlocking yoke drops off the cam, it is propelled back to its starting point by the unlocking yoke spring for the date indicator.

The jumper and jumper spring ensure that only one tooth advances, and that the date is held in the display for twenty-four hours.

Dismantling
1. Make sure there is no tension on the unlocking yoke for the date indicator. The best way to do this is to remove the spring.
2. Slacken the two screws holding the date indicator guard, but slacken the screw nearer to the date jumper just a little more than the other screw.
3. Lift the date indicator sufficiently for the jumper to slip under it.
4. Remove the two screws holding the date indicator guard, and the guard itself.
5. Remove the date indicator.
6. Hold the crook of the date jumper spring lightly with your fingernail while you lift the jumper spring out of the plate.

7. Remove the jumper.
8. Remove the calendar driving wheel and its holding screw.
9. Remove the unlocking yoke for the date indicator and the hour wheel.

With the exception of the jumper spring and date indicator, all components may be put through the cleaning machine.

Reassembly
1. Replace the hour wheel, unlocking yoke, the calendar driving wheel with its screw, and the date jumper. Remember to lubricate with watch grease as you go.
2. Replace the jumper spring, holding it in position with your fingernail while you load the spring behind the jumper.
3. Replace the date indicator, allowing it to rest on the jumper.
4. Position the date indicator guard and screw in its two screws, but only so that the guard is held very lightly.
5. Hold the guard in contact with the date jumper as you pull the jumper back to load the jumper spring, and allow the jumper to engage between two teeth of the date indicator. As you do this, make sure the pawl in the unlocking yoke doesn't foul teeth in the date indicator.
6. Tighten the screws in the date indicator guard.
7. Replace the unlocking yoke spring, remembering to load it behind the unlocking yoke, and to grease the point of contact.

When a watch has both day and date, the day indicator might be held in position by a gib; on some watches this lifts up over the hour wheel, or it may slide back out of a groove in the hour wheel. Be careful to identify which you are working on. The principle of operation is similar to that already described.

5 PART JOBS

Not all watch work involves an overhaul. When a watch comes into a shop for, say, glass only, or mainspring only, or button and stem only, it is sometimes known as a 'part job' to distinguish it from a full overhaul, and is therefore likely to be fitted around more major work. The common part jobs not already detailed include selecting and fitting mainsprings, buttons and stems, watch glasses, spring bars and testing for water resistance. These part jobs could of course be in addition to an overhaul. (Fitting balance staffs is not included because in contemporary horology it is becoming more common to fit complete balances. For those who wish to know about balance staff fitting and turning, it is dealt with at length in *The Clock Repairer's Manual* by the same author.) We shall also discuss movement identification, and how to order materials.

MOVEMENT IDENTIFICATION

For many part jobs a prerequisite is to be able to identify watch movements. There are many factories throughout the world producing watches: some put their own name on the dial – for instance Rolex, Omega and Seiko – whilst others produce watches which carry a dial name that is not that of the manufacturer.

When material is required for a watch, it becomes necessary to identify the manufacturer and the model. This is done by finding what is known as the calibre and the calibre number. Not always, but often, the manufacturer marks the movement somewhere with a name or a symbol and a number which will positively identify it. The most likely places to find the calibre might be:

(a) on the inside of the bottom plate under the balance;
(b) on the outside of the bottom plate near the barrel arbor hole;
(c) on the barrel or train bridge;
(d) on the inside of the bottom plate under the escape wheel.

On occasions the calibre and number are in two different places.

Not all watches are marked with a calibre or number, in which case the movement may be identifiable from one or more of the many identification systems that exist; alternatively your materials dealer may be able to identify it, since he/she will probably have access to identification catalogues not generally available. These systems may also refer to parts that are interchangeable between movements, even by different makers; they include: *Bestfit Encyclopaedia of Watch Material Part 1 & 2, the Flume System Band 2/3 and K3, The Official Catalogue of Swiss Watch Repair Parts, Ebauches SA Catalogue*, and catalogues by makers including Seiko, Citizen and Ronda. This list is not comprehensive, as there are many other manufacturers who provide catalogues that identify not only watches and parts, but also parts that are interchangeable between calibres; they sometimes also include lubrication details. Some of the catalogues may be out of print, but occasionally they are passed from one repairer to another, or they may be bought at auction, or they may be available through specialist booksellers. (See the address section which includes addresses of material houses and a reference for obtaining other horological services.)

The table on page 67 is a working guide for ordering parts and materials; it also gives the

Table

10	Batteries	Battery reference number
1	Balance complete	Calibre no; whether shock proof and type; beat device or no beat device; catalogue reference number, e.g. Ronda no
3	Balance staffs	Calibre no; ordinary or shock system; catalogue ref no
1	Barrel arbor	Calibre no
1	Barrel complete	Calibre no
1	Beat set device	Calibre no; shock system
6	Buttons	Chrome/gold plated; tap/diam; ord/dust proof/water resistant; short/long pipe
1	Cannon pinion	Calibre no; part no to distinguish type; overall height
1	Centre wheel	Calibre no; state centre seconds (c/s) or non c/s
1	Click	Calibre no
1	Click spring	Calibre no
1	Clutch wheel	Calibre no
1	Crown wheel	Calibre no; part no
1	Crown wheel ring	Calibre no
3	Cap jewels	Diameter in hundredths of mm
2	Cap jewels Incabloc	Inca ref no
1	Escape wheel	Cal no; part no to establish pivot type (straight or conical)
1	Fourth wheel	Cal no; part no to establish pivot type
3	Glasses	Make; size; low dome/high dome/armoured ring and colour
1	Hour wheel	Calibre; part no; c/s or non c/s; overall height; calendar
3	Jewel holes	Convex or flat; hole size and diam
2	Jewel holes shock	System; system ref no
1	Minute wheel	Calibre
3	Mainsprings	Calibre no; mainspring catalogue ref no
1	Pallet complete	Calibre no; part no or pivot description, straight or conical
1	Ratchet wheel	Calibre no
1	Regulator	Cal no; type of shock or non shock; whether beat device
1	Roller	Cal non; type of shock or non shock
1	Setting lever	Calibre no
1	Setting lever spring	Calibre no
1	Setting wheel	Calibre no
5	Shock springs	System with ref no or state calibre used on
3	Stems	Calibre no or catalogue ref no; state if split stem
1	Third wheel	Calibre no; state type of pivots
1	Winding pinion	Calibre no
1	Yoke	Calibre no
1	Yoke spring	Calibre no
1	Centre seconds wheel	Calibre no; overall height

most usual quantity to order, and the information required by material houses.

SELECTING AND FITTING WATCH MAINSPRINGS

Carbon steel mainsprings are rarely fitted today: they have been replaced by the unbreakable (U/B) mainspring, easily distinguishable by its white colour and 'S' shape when it is removed from the barrel. Different manufacturers claim all, or some of the following characteristics for their U/B mainsprings:

unbreakable	self-lubricating
non-setting	they have a sharpened hook
rust-resisting	they are sold in hermetically
antimagnetic	sealed packets

The mainsprings are supplied in a metal ring so they will push straight out and directly into the barrel; they do not have to be wound into a mainspring winding tool first. A new U/B mainspring should require no lubrication; however, repairers that I have asked all say they lubricate U/B mainsprings unless they are new – and so do I.

The only two drawbacks I have found with U/B mainsprings is that if one does break – and they sometimes do at the outer end – it is impractical to repair it; the other is that on occasions, the inner coil of the mainspring is too large for the barrel arbor so that it cannot be closed.

Mainspring Ends
Whilst mainspring inner ends are always 'eye' end, outer ends can vary; but all suppliers are likely to include a riveted or spot-welded outer end, an eye end, a 'T' end and a slipping end for automatic watches. An example of each is shown in Fig. 79.

Mainspring Sizes
Different ways and units are used for measuring mainsprings. For U/B mainsprings, metric measurement is used. We are interested in mainspring height (H), in force or thickness (F), and in length (L): and they are always given in that order. We also need to know the inside diameter of the barrel (Ï); sometimes this is inserted between the height and force, at other times it follows the length.

In the days of carbon steel mainsprings (pre-1950s) there were certain proportions for the inside diameter of the barrel relative to the barrel arbor and main spring; these were (a) the space inside the barrel, which would be occupied by one third mainspring, one third space and one third arbor; (b) the mainspring thickness which would be between $\frac{1}{30}$ and $\frac{1}{32}$ the barrel arbor diameter; (c) the force of the mainspring, which would be $\frac{1}{100}$ of the inside diameter of the barrel. In contemporary horology, these proportions are unreliable, as watch design has changed. For example, mainsprings are often longer and weaker than the old proportions would dictate, to give a more uniform power output and thus more uniform amplitude to the balance.

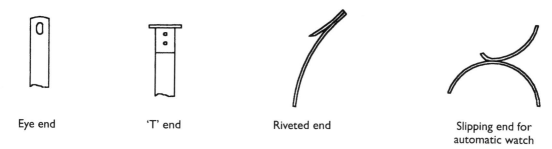

Eye end   'T' end   Riveted end   Slipping end for automatic watch

Fig. 79 Common mainspring ends.

Ideally, when a new mainspring has to be fitted, look up the calibre of the watch in a mainspring catalogue and select a replacement mainspring by reference number. Even then, it is a good idea to compare the old and new mainspring for size, and to check that both are indeed correct for the watch.

Checking Height

Barrels may have a flat cover or a raised one (Fig. 80). The correct height of the mainspring can be assessed by measuring the height between the cover and the bottom of the barrel, and choosing a height at least 0.05mm less than your measurement. Watch mainspring heights rise in 0.05mm steps for the smaller sizes, and 0.1mm for the larger. You would be fairly safe in coming down two heights if necessary – probably more if the size you need is no longer available.

Mainspring Length

The length of the mainspring is likely to be dictated by the force and inside diameter of the barrel, but as already mentioned, old rules of thumb are not infallible. The length can be varied by about 10 per cent safely. It is a fallacy to think that a longer mainspring will cause the watch to go for longer between winds: the mainspring needs space, and an over-full barrel will drive a watch for a shorter period, not longer.

When having to select the nearest most suitable mainspring when the ideal one is not available, take the following into account:

(a) to halve the height of a mainspring is to halve the strength;

(b) to halve its length is to double its strength;
(c) to double its thickness is to make it eight times stronger;
(d) An old, well worn, low-grade watch, perhaps with a pin pallet escapement, will probably take a stronger spring rather than a weaker one;
(f) If too strong a spring is fitted, there is a risk of the impulse pin striking the wrong side of the notch due to too high an amplitude. If this happens, fit a weaker spring;
(g) If the amplitude of a watch is too low, don't automatically fit a stronger spring, but look for the actual fault. If it does happen to be too weak a spring, then – and only then – fit a stronger one.

A Practical Example

Suppose you needed a mainspring for a watch that you couldn't identify, and you feel the old spring may or may not be the correct one; you might proceed as follows:

1. Establish whether the barrel cover is flush or raised.
2. With a metric micrometer, measure the height from the barrel cover lip to the outside of the barrel, and the thickness of the bottom of the barrel. Here we will assume a flat barrel cover and that heights were 2.77mm and 0.28mm. (Fig. 81.)
3. Taking 0.28mm from 2.77mm, we arrive at 2.49mm as the height of the inside cover of the barrel. To allow some clearance between

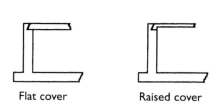

Flat cover   Raised cover

Fig. 80 Barrels with flat and raised covers.

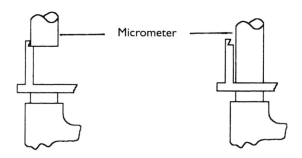

Micrometer

Fig. 81 Measuring the height of the barrel wall.

the mainspring and the barrel wall we need a height of less than 2.49mm. The mainspring catalogue shows us that the nearest height below 2.49mm is 240 (2.4mm).

4. Now measure the inside diameter of the barrel to the nearest half millimetre. Here we will assume a measurement of 22mm. This means that I am looking for a mainspring H240 Ø22 F0.22.

5. Referring to my catalogue, the nearest mainspring is an H240 Ø22.5 F0.24 and L820, but my supplier doesn't keep the full range of mainsprings, just the most popular sizes. The nearest he stocks is H220 Ø21.5 F0.24 L680 which I happen to have. I fit this stronger mainspring, and when fully wound, the balance has a healthy amplitude.

The slightly lower height helped offset the greater thickness and was aided by the fact that it was an old medium grade watch that was past its prime.

SELECTING AND FITTING BUTTONS AND STEMS

Stems

Watch stems are always made of hardened and tempered steel. They are selected according to the calibre of the watch, and are normally supplied in packets of three. Each stem has a long metric thread which is cut to length at the time of fitting with stout top cutters. Though hardened and tempered, stems threads are easily cut this way, and the cut end is then dressed with a fine file.

By far the most popular thread size is tap 9, which means that the outside diameter of the thread measures approximately 0.9mm. Until the advent of quartz watches, it was a very rare occurrence indeed to come across a tap 8 (0.8mm diameter), although taps 10, 11 and larger were fairly common. With the advent of quartz watches, there are stems with thread sizes as low as tap 7.

Fig. 82 shows five fundamental shapes for stems.

Stem Extenders

When a stem breaks, it is usual to fit a new one – but often there is an alternative solution, namely to fit a stem extender. These have a normal thread to take a button, but instead of a main bearing, square and small pivot, they simply have a female end which screws onto the broken stem. Different tap sizes are an option, with a choice from tap 6 to tap 12. Even so, an extender can only be used where there is sufficient space between the movement and the case, and enough thread left on the original stem to take one. (Fig. 83)

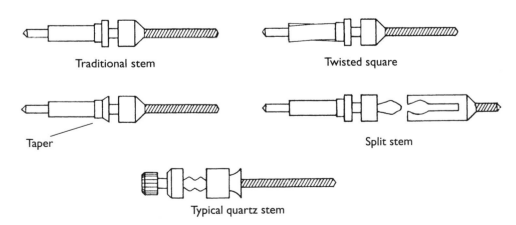

Traditional stem

Twisted square

Taper

Split stem

Typical quartz stem

Fig. 82 Five fundamental stem shapes.

Fig. 83 Stem extender.

Buttons

The appearance of a watch button, often called a crown, will either match or contrast with the watch case, depending upon the design of the case. Watch cases can be chrome, stainless steel, silver, gilt, gold plated, rolled gold, 9 carat (ct) gold, 18ct gold, white gold or platinum, to mention just those more commonly encountered. In repair work we rationalize what would otherwise be a massive investment in watch buttons by stocking a limited range that would suit most watches, in addition to a few 'special' buttons to suit high-grade watches such as Rolex, Omega and one or two others. It is also usual to stock some of the buttons that have less traditional characteristics, but which often occur nevertheless.

Buttons are nearly always either white or yellow in colour, and many of the workshops I have worked in stock chrome buttons for the more common watches that have chrome or stainless-steel cases, and rolled-gold or perhaps gold-plated buttons for watches that are gilt, gold-plated, rolled gold, 9ct or 18ct gold.

Button Sizes
The thread of a button is always metric to match the stem. Buttons are also sized by outside diameter, which can be in millimetres but is much more likely to be in douziemes. A douzieme is a continental unit of measurement approximately equal to 0.19mm. For water-resistant and waterproof watches a third size needs to be known, namely the outside diameter of the pipe that protrudes from the case, so that a water-resistant seal can be made. This pipe is sometimes referred to as the 'tube' or the 'pendant', or even the 'pendant tube'. Diameters 200 and 250 (2mm and 2.5mm) are the most

common tube sizes, though other sizes do exist.

Button sizes vary, but diameters 22 and 24 are most common for ladies' mechanical wrist watches, with diameters 26 and 28 common for gents' mechanical watches. (Most of the time we don't write or say 'douziemes', but it is assumed.) For a variety of reasons the buttons of quartz watches are often much smaller in diameter, perhaps measuring as little as 13 douziemes.

Button Types
There are three button types: ordinary, dust-proof (D/P) and water-resistant. Before the Trades Description Act, the term 'waterproof' was used much more liberally than it is now, and buttons that used to be called 'waterproof' are now known as 'water resistant'. Water resistance is discussed on p. 77.

Ordinary buttons are fitted where the case is rounded in two planes or – and this is particularly the case in older watches – where there is a short pipe (not to be confused with a tube to make the watch water-resistant) protruding from the case. The button is hollowed to accommodate a rounded case or/and a short pipe. (Fig. 84)

A dustproof button does what its name suggests: it helps to keep dust out of the watch. It has a secure spring bush which maintains contact with the case whether it is winding or hand-setting. The case is flat where the dust-proofing bush makes contact. (Fig. 85)

Fig. 84 An ordinary button is fitted to a case that is rounded in two planes or has a small pipe.

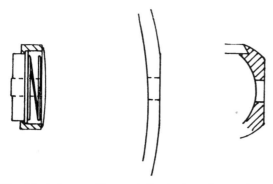

Fig. 85 A dustproof button is fitted where the case is machined with a flat at the 3 o'clock position.

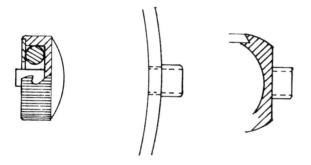

Fig. 86 A water-resistant button is fitted to a water-resistant case with a pipe or tube.

The most common type of water-resistant button is bulkier than the other two types and has an 'O' ring inside to make the watch water-resistant. The outside of the 'O' ring presses against the inside of the button, while the inside of the 'O' ring presses against the outside of the tube in the case. (Fig. 86) The tube is lubricated with silicone grease before the button and stem are inserted into the case. There is also another type of button with an 'O' ring positioned part-way down an extended pipe intended to engage with the inside of the tube in the case.

Long Pipe Buttons

When a stem breaks, very often it does so just where the stem enters the button. This means that a button with a longer-than-usual pipe may be fitted instead of a button and stem. Such buttons are known as 'long pipe buttons', and they have a shallow thread to make up for the missing bit of stem. It is not usual to stock a comprehensive range of long pipe buttons, but simply a limited range of the most popular sizes. Materials catalogues may also show something called 'button boots' or 'crown attachments': these are similar to stem extenders, the main difference being that they have a much shorter thread on the male part, thus converting the button into a long pipe.

Fitting Buttons and Stems

1. Start by identifying the movement by calibre and number.
2. Select a new stem – this may have come from stock, or it may have been specially ordered – and compare it with the old one just to confirm that it is correct in its requirements; also establish the thread size, if necessary by measuring the outside diameter of the stem with a metric micrometer.
3. Select a suitable button according to the flow chart below.
4. Hold the new stem in a pin chuck, and screw the stem into the button. This is not essential, but it does ensure that both threads are clear so that if difficulty is encountered later when screwing the cut stem into the button, it will be the stem that has a burr and not the button.
5. Remove the button from the stem, and with the movement held centrally in the case, put the stem into the movement, tack down the setting lever and pull the stem in and out to confirm that it is properly located. Leave the stem in the winding mode for cutting to length.

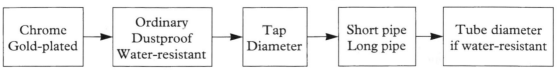

6. If the watch takes an ordinary or a dustproof button, cut to length, leaving four good threads protruding from the case. If the watch is water resistant and has a pipe protruding, cut the stem flush with the end of the pipe. In the vast majority of instances I have found that by cutting where recommended, there is a minimal amount still to remove from the stem for the button to be in the right place relative to the case.

7. Remove the stem and file the cut end square with a fine file such as a 6in No. 4 double-cut file; also remove the burrs. For this I find it better to hold the stem in my hand, though you could hold the stem in a pin chuck by its largest diameter, the main bearing, but still try to grip the thread with your fingernails: it helps to prevent breaking the stem. As a third alternative, grind the stem with a small, fine grindstone held in a watchmaker's lathe. Hold the stem by the main bearing in a pin chuck, but with the other hand, hold the thread of the stem with tweezers. This will minimize the chance of the stem breaking.

   *Never leave the threaded end of a stem pointed in case it raises a pip in the end of the button: always square the end and de-burr.

8. With the stem still held in the pin chuck, screw the stem into the button, but hold the stem by the thread in brass-lined pliers for the final tightening to avoid breaking the stem. Screw the button on 'squeaky tight' (you will sometimes hear a squeak during tightening).

9. Try the button and stem in the watch to see how well the button is positioned with respect to the case. With the movement centralized and the button and stem pushed into the wind position, the button should be just clear of the case. A distance of, say, 0.05mm to 0.1mm is about right. The button mustn't touch the case, yet to be too far off would look ugly, and the stem would be more likely to break if knocked.

10. It may be necessary to trim a little off the length of the stem. Try to judge how much to take off: if necessary use the cut piece of stem to measure with. (Units such as 'one-and-a-bit threads' are quite adequate for this measurement.)

    *Do remember to tighten the button on the stem fully, otherwise over a period of time the button will creep further onto the stem until it touches the case – this could even draw the button away from the movement, preventing the full engagement of the clutch wheel and winding pinion, leading to worn parts.

11. When replacing the button and stem for the last time, remember to lubricate the stem with watch grease, and if the case is water resistant, lubricate the tube with silicone grease as well.

If the job of button and stem fitting is done properly, even with a tight cannon pinion, the button should not unscrew from the stem during hand-setting. Only under special circumstances would I use Loctite Studlock 601 to secure a button to a stem – usually a stem is too short and there is no replacement available. At times I have fitted a brass plug at the bottom of a button to make up for a short stem, but only if I am satisfied that a competent repair can be effected and that there is sufficient thread to hold the button securely and true.

SELECTING AND FITTING WATCH GLASSES

For some years now it has been unusual to fit 'glass' glasses to watches: instead we fit unbreakable acrylic glasses (U/Bs), or mineral glasses. Even the very thin 'Hunter' glasses that were available in the 1970s have been replaced by unbreakable ones. Material houses often stock more than one make of watch glass, and can generally supply a full range of glasses measuring from about 12mm, known as a 120 (one-twenty), up to pocket-watch size of 500. Until about the

| Low dome | High dome | Tension ring |

Fig. 87 Three types of unbreakable glass.

Fig. 88 The bezel is machined with an undercut to take a low- or a high-dome glass.

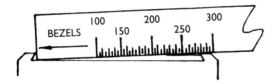

Fig. 89 Glass-measuring tool here reading 278.

1970s, the three most popular types of round U/Bs were low dome, high dome and tension ring, also known as armoured ring. (Fig. 87)

Low Dome Glasses

These are fitted to watches where height is not important. They include watches without a seconds hand, and watches with a small seconds hand. A watch with an ordinary low dome glass would not be considered water-resistant. The dial is likely to be flat, and the bezel will be machined, with an undercut to take the glass. (Fig. 88.)

High Dome Glasses

These are fitted where extra height is necessary, for example where centre seconds are used, or where the dial is raised or has batons that are raised up from the dial. Again, the bezel is undercut, and such watch cases will probably not be considered water-resistant.

Tension-Ring Glasses

These are sufficiently high to be used with centre seconds. The bezel will have been machined vertically to accept the glass, which is sandwiched between the bezel and tension ring. These glasses have the potential to be water-resistant and are obtainable with chrome and gilt rings.

Sizes

Low and high dome glasses are usually ordered in threes, and sizes rise in 0.2mm. An experienced worker will notice that the most frequently used sizes end in a six or an eight: for example 276 or 278; 286 or 288. Tension-ring glass sizes rise in 0.1mm, between 180 and 349.

Measuring Bezels and Glasses

A vernier calliper could be used, and is certainly better for tension-ring glasses, but a simple glass-measuring tool is available through material houses. The most popular tool is calibrated to 0.5mm, and although glasses rise in 0.2mm, the tool is quite adequate for the job. (Fig. 89)

Fitting Glasses

Low and high dome glasses are fitted by dropping the bezel or case over a 'dolly', placing the glass on the dolly and using a second dolly to compress the glass, thus reducing the diameter. When the tool is released, the glass should be held securely so that it will not turn, but nevertheless is not too tight; glasses that are over-compressed tend to crack over a period of weeks. (Figs. 90 and 91)

Ideally, tension-ring glasses are fitted with a stake on the inside of the case as support, as the glass is pressed in with a pusher that applies pressure to the outside edge of the glass. In the absence of a stake, screw on the case back to give the case support, and then press the glass into place.

When you first start fitting tension-ring glasses, you might find it difficult to judge how much pressure should be used, and will probably break a few. If you can *almost* push in a glass with your fingers, the correct size is likely to be

Fig. 90 Fitting a glass with the Robur glass-fitting tool.

0.1mm or 0.2mm larger. Sometimes it helps to fit a glass if its leading edge is rounded with a buff stick to give it a lead in. It also often helps to put a little silicone grease around the outside corner of the glass.

Don't be tempted to use the old tension ring in a new glass – and certainly don't fit a high or low dome where a tension ring is called for, as it will not be held securely. If the glass does come out, the hands and dial will almost certainly be damaged.

Shaped and Special Glasses

It is unreasonable to expect a repairer to be intimate with the vast range of special glasses, or to have the special equipment to fit all glasses, or to have a totally comprehensive glass stock; and personally I have always favoured sending 'specials' to a specialist glass fitter for a truly competent job. Up to the present time I have also sent watches requiring mineral glasses to a specialist fitter, though many repairers nowadays in fact stock and fit them themselves; for instance, Sternkreuz offer a good range of mineral glasses through material houses, in a variety of thickness and diameter. Standard glass

*Fig. 91
Checking that the
glass is secure.*

thickness is between 1.0 and 1.1mm, thin glasses are between 0.7 and 0.8mm and thick glasses are available between 2.4 and 2.6mm. Some cases have a gasket between the case and the glass; there are both 'L' and 'I'-shaped gaskets, and these are also readily available. A range of adhesives can be used with mineral glasses, though some need UV to cure. Should you decide to send bezels and glasses to a specialist, remember to include any case numbers and dial names as the right glass might be available off the shelf.

Glass Removal

Low dome, high dome and tension-ring glasses are pushed out from the inside of the case with thumb pressure. If the top of a tension-ring glass is broken away, get a small screwdriver and shape it to a knife edge on an Arkansas stone. If you then shave away the glass in just one area this will reduce the pressure and the remaining glass will 'peel' out.

Although a normal glass-fitting tool can be used for removing glasses, low and high domes may be removed and replaced using a Vigor Crystal Lift. A significant advantage is that the back of the watch doesn't need removing, and the movement need not be removed from the case. This tool pays for itself quickly and a busy workshop would find it a great asset. The tool has a number of claws and these compress the glass for easy removal and likewise replacement. (Fig. 92)

SPRING BARS

There are about twelve different types of spring bar for attaching bracelets or straps to a watch: these include universal, telescopic, simple, thin, flanged, curved, suvoy, extra long, female, stainless steel and ultra-thin, and you can buy a set of spring bars for bracelet clasps and for attaching buckles to leather straps. I have never found it

Fig. 92 Vigor Crystal Lift.

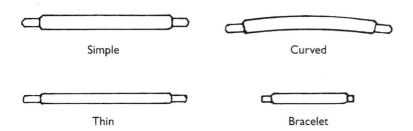

Simple

Curved

Thin

Bracelet

Fig. 93 Four different types of spring bar.

necessary to stock a fully comprehensive range of spring bars, but instead stock a full range of simple bars, a very small range of curved bars, a small range of thin bars and a set of bracelet bars. This has covered all my needs over many years. I order spring bars by the dozen, and find that 18mm simple bars are needed more than any other. (Fig. 93)

Broadly speaking, lengths range from 4mm to 35mm, rising in 1mm steps with diameters from 1.3mm for an ultra-thin bar, to 2mm for extra long bars.

To select a spring bar, first measure the gap between the lugs of the watch case, and then select the appropriate spring bar; for instance a measurement of 18mm will accept an 18mm bar. A compressed simple spring bar will measure about 17.5mm so it can be inserted between the lugs of the watch. (Fig. 94)

Although it may seem a good idea to stock universal spring bars, to save keeping so many sizes, I for one would never fit a universal bar to a watch that has a leather strap: the strap can grip the bar, and sideways movement of the strap can pull the lug out. I once very nearly lost my own watch in this way – luckily someone saw it

fall from my wrist and handed it back to me. I have never fitted universal bars since, although they would be safe with bracelets. I therefore always try to fit simple bars – but once fitted, unless the lug has a 'through' hole to push the bar out, the bar will need to be broken out or cut with top cutters. A special tool is available for inserting and removing spring bars, but tweezers are quite adequate. Simply hold the sprung end above the shoulder, put the other end of the bar into its hole, then compress the end and insert it into the hole in the lug. To remove, insert a screwdriver blade between the watch lug and the shoulder of the spring bar, compress the sprung end and pull it away from the hole. The use of safety glasses is worthwhile when fitting bars as sometimes the sprung part or the whole bar does fly.

WATER RESISTANCE

There are two common origins of rust in a watch: one is condensation perhaps from moisture in the air, the other is water entering the watch.

Condensation
Watches breathe to a greater or lesser extent, and because they do, the moisture content of the air can be quite high. While the watch is warm, the moisture is held in the air; however, when the watch is subjected to cold, the air can no longer hold the same amount of water, and moisture is deposited as condensation on the inside of the watch glass.

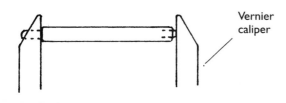

Vernier caliper

Fig. 94 Measuring a spring bar.

If the watch is worn, its back is comparatively warm and little, if any, condensation would be evident inside the case back. The exposed areas of the watch are the first to get cold, such as the glass for example, which explains why condensation can appear on the glass yet apparently no water has entered the case. Eventually the hands, dial and bottom plate can suffer with condensation, which is why the mechanism under the dial is one of the first to suffer rust.

Sometimes it is possible to get rid of condensation by exposing the inside of the case and the movement to the warmth of a bench light and, provided it isn't a particularly humid day, after casing there is no more problem. If, however, the condensation is heavy, the watch should be stripped and cleaned before rust sets in.

The risk of condensation is higher for people who have their hands in hot water frequently – for example medical personnel – and also for those who work outside in extremes of weather. For such people a water-resistant watch would minimize the risk of water getting into the watch.

Unbreakable watch glasses are hygroscopic, so even if a watch is sealed in dehumidified air, eventually the outside air and its moisture content will pass through the glass. Modern watch glasses – for example mineral glasses – are less likely to suffer from this.

Water-Resistant Cases
In a three-part watch case, water can enter a watch between the glass and bezel, the bezel and case, the case and the watch back and between the case and button. Today, however, a lot of watch cases are two-piece, which straightaway reduces the number of areas where water can ingress; and some cases are essentially one-piece, leaving only the glass and button where water may penetrate.

Protecting the Bezel and Glass from Water
Water getting in between the bezel and glass is minimized by careful machining of the groove in the bezel that takes the glass, but also by having the glass sandwiched between the bezel and a metal ring. The bezel is machined with a vertical wall to take the glass, and the glass, with its armoured ring, is forced into place.

Preventing Water getting past the Button
The case has a protruding tube, through which the stem is passed to wind the watch and set the hands. Water could get in between the tube and the case and, where I suspect this might be happening, in the past I have tinned the tube where it pushes into the case, using soft solder to make a watertight seal. To prevent water getting past the button, a gland is compressed between the tube and the inside of the button; this used to be rubber but is now more likely to be some other material.

The Case Back
A common and traditional way of preventing water entering between the case back and the case is to have a threaded case back rather than a snap-on back. The case is also threaded, and provision is made for a gasket or 'O' ring to make a water-resistant seal between the case and the back. (Fig. 95)

Both gaskets and 'O' rings are selected by outside and inside diameter, and it is possible to get gaskets of different thickness. Both should be lubricated with silicone grease before fitting, to prevent the seal from moving as the case back is tightened. Gaskets and 'O' rings are available from material houses in mixed assortments, or graded. I prefer graded, as replacements are ordered by size and I always try to keep a full range; if you stock assortments you often find

Fig. 95(a) A water-resistant case with a gasket. (b) A water-resistant case with an 'O' ring.

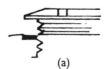

(a)

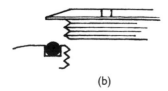

(b)

you quickly use up all the popular sizes and are left with the unpopular ones.

Testing for Water Resistance

There are differing views on the most reliable way of testing that a watch is water-resistant. Here we will be looking at two ways of testing: one uses water and is sometimes called a wet test, where we will test to 3 bar; the other is a dry test using air pressure alone, which here will be to 10 bar. It is possible to test beyond 10 bar but you will be limited by your equipment. When testing we refer to the number of bar or the number of atmospheres or testing to a particular depth in feet or meters. One bar is equal to 10 Newton per sq. metre. For practical purposes one bar is also approximately equal to one atmosphere or 33.3 feet depth or 10m depth.

A watch that is marked water-resistant should withstand splashes, sweat, raindrops and similar, but should be wiped off as soon as possible after being in contact with moisture. The watch need not be tested to any particular depth. Watches marked water-resistant to a particular depth should be tested to that depth. As a guide, for swimming the watch should pass a 50m or 5 bar test, and for snorkelling and general water sports a 100m test or 10 bar test. For scuba diving and for professional deep-sea diving a much more substantial test must be passed. Inexperienced, untrained repairers are discouraged from testing beyond 10 atmospheres because the life of a diver may depend on a reliable test, and this book does not cover all you need to know about making a diver's watch safe for diving. A diver's watch should be serviced and sealed by a specialist, probably to ISO standard.

Often the extent of water resistance is marked on the watch dial or watch case. If you have no testing facilities, before you accept a watch for repair, morally you should advise your customer that you cannot test for water resistance. If a watch case continues to fail a test, again the customer should be advised. Probably it is not very practical to give a modern quartz watch a wet test every time a battery is fitted, and the customer should be told that there is no guarantee that the watch is water-resistant. A dry test, on the other hand, is done in seconds, but should be charged for as the test equipment is costly.

Wet Test using Water (Bergeon 555)

Many consider a wet test to be the most reliable test of water resistance. The watch is prepared by replacing the old glass with a new one, by fitting a new seal to the back and tightening the back securely and fitting a new button to the tube. The movement is left out of the case, and to prevent the button from blowing off the case, a wide elastic band is passed around the case to hold the button in place.

The tank is part filled with water (distilled ideally) and the prepared watch is suspended over the water, but definitely not in it, and the tank is sealed. Air to a maximum of 3 atmospheres is now pumped into the top of the chamber and left for three minutes so it has a chance to enter the watch. On no account remove the lid of the tank without first releasing the compressed air through the valve.

After three minutes, the watch is immersed into the water and the compressed air is slowly released through a valve. If compressed air had entered the case, it will now escape and air bubbles will be seen to rise from the watch at the point where air entered. If this happens, the watch case has failed the test and the watch should be removed from the water while the tank is decompressed completely. If no air escapes, the watch is considered water-resistant at least to the depth to which the pressure was raised. If air bubbles are seen to come from between the bezel and glass, a new glass will be fitted. If the problem persists, the bezel would be put into the lathe and turned to ensure the aperture is round and has a good smooth finish. If bubbles emerge from the button, a new button would be fitted or a new tube. If air escaped from the case back, a new seal would be fitted. Look for a steady stream of bubbles, as quite often one or two bubbles can rise due to air being trapped in the hole for the spring lug or other places. (Fig. 96)

Fig. 96 *Testing for water resistance with the Bergeon 555 tester.*

Before testing the Sigma, it is set to between 0.5 bar and 10 bar, and the duration of the test to between 30 and 90 seconds. The longer the test, the more valid the result. It is also possible to set the Sigma to do a low pressure test of 0.5 bar, followed by a higher test of up to 10 bar. Some repairers are of the view that a single test at high pressure is no guarantee that the watch will be water-resistant at low, because the pressure itself can help to close a seal thus keeping water out. This combination test is therefore reassuring. The tests are pass or fail, with light-emitting diodes showing which. The display indicates the pressure as it rises. (Fig. 97)

To complete the process, finish decompressing the tank then remove the case, dry it and insert the movement. If you are brave enough, the test should be repeated with the movement inside. In theory, with this test, no water should enter the case, even if it is not water-resistant.

Water Resistance Test using the Sigma SM 8800 A

This test does not involve the use of water at all, and so there is no risk of water entering the watch.

Fig. 97 *Dry-test with the Sigma SM 8800 A.*

Full instructions are not given here – they come with the Sigma – but you would first set the pressure to between 0.5 bar and 10 bar and the duration of the test to between 30 and 90 seconds. The watch, with its strap or bracelet, is placed on the sensor table and the top of the test chamber is closed. That action starts the process. On no account unlock the test chamber until the test is completed and the chamber is de-compressed.

The table rises, and both 'strain monitor' and 'pressure select' zero then rise until the selected pressure is reached (here, 10 bar.) As this happens, the strain monitor will display a value of between 400 and 600 microns, and then the chamber will stabilize (one micron is equal to 0.001mm). Shortly after the strain monitor resets to zero, and the actual test starts. The strain monitor displays a deflection value in tenths of a micron (on my test watch, 8.5). Next, the chamber exhausts through the back of the Sigma, and at the same time indicates a pass or fail on a green or red LED. Only after exhaustion and when the pressure select has returned to zero, and the pressure resets to the test value, is it safe to open the chamber.

6 STOP WATCHES

A stop watch has the facility for short interval timing but does not show the time of day. There is a seconds-recording hand, and a hand for recording minutes, and these can be stopped, returned to zero and started again by a single pusher (the button) or by a slide and pusher. In the latter, the slide starts and stops the watch while the pusher returns the hands to zero, and the advantage with this arrangement is that timing can be interrupted and recommenced without returning to zero, thus avoiding the need to add a number of intervals together. Many early watches had an 18,000 train (an expression to indicate that the balance vibrated at 18,000 vibrations an hour for the watch to keep exact time), so the dial was calibrated to read $\frac{1}{5}$ seconds. This is because 18,000 vibrations/hour is equal to five vibrations or beats each second. There are now much faster trains, for example 36,000, so mechanical stop watches reading $\frac{1}{10}$ second are available.

Dials can be quite simple, as with the watch we will be working on, or may be specially marked for a particular use. For example, a stop-watch dial designed specifically for the use of a football referee and one for a yachtsman are quite different. Our watch is quite commonplace and could be used for a school sports day or for timing exams. It records $\frac{1}{5}$ seconds up to 30 minutes, and starts, stops and returns to zero with a single pusher, which is also the winding button. (Fig. 98)

How a Stop Watch Works
Referring to Fig. 99, in the centre right of the picture is a pillar wheel with five posts (the pillars) and fifteen teeth at the base. For single pusher operation there will always be three times the number of teeth to the number of pillars. The pillar wheel is held down by a left-hand thread screw; this is because it turns in an anti-clockwise direction and so will not tend to unscrew in operation. The pillar wheel is held in the right position to operate the levers by a jumper under the influence of a jumper spring. When the button is pushed, an operating lever advances the pillar wheel by one tooth.

Fig. 98 A one-fifth second stop watch with single pusher.

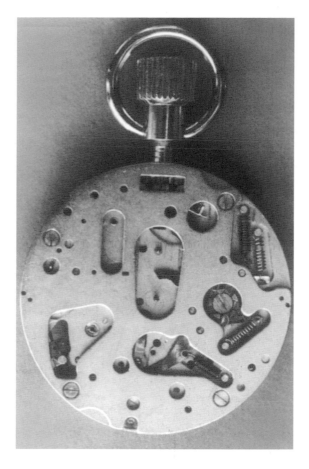

Fig. 99 The stop work exposed showing the pillar wheel.

The hammer for the seconds-recording heart is raised away from the heart with its 'beak' resting on a pillar of the pillar wheel. The minute-recording hammer is raised by the seconds-recording hammer. Having two separate hammers is useful because, when operated, both hands zero correctly even if one face becomes slightly worn or a pivot hole becomes worn. When both hammers are on one lever, the acting face of the minutes-recording hammer may need to be dressed so that, when returned to zero, any slackness would be on the minutes-recording heart, and not on the seconds-recording heart. (See p. 98.) Slight play on a seconds recording heart could account for ⅕ second error or more, whereas a similar play in the minutes-recording heart would barely be noticed and the correct minute would still be easily read.

Stop Mode

When the pillar wheel is turned by one tooth, one end of the whip-balance stop is raised by a pillar, and this causes two things to happen: (a) the whip-balance stop interferes with, and stops the balance; (b) the middle of the arm of the whip-balance stop presses on the blocking lever, which in turn lowers the blocking lever onto the seconds-recording wheel. The hammers remain raised, as the seconds hammer is still riding on a pillar. (Fig. 100)

Return to Zero

When the pillar wheel is advanced one more tooth, the whip-balance stop continues to hold the balance and press on the blocking lever. The seconds hammer falls into a space between two pillars, allowing the minutes-recording hammer to fall also. The hammers strike the periphery of their respective hearts, and it matters not where they strike because they will always return the hearts to the zero position. (Fig. 101)

The Hearts

Fig. 102 shows the design of the heart, the seconds-recording heart being struck by its hammer and the zero position.

Start Mode

With the watch in the 'start' mode, the 'whip-balance stop' is clear of the balance while the other end lies between two pillars of the pillar wheel. As the whip-balance stop is operated, the balance is given a slight turn to help it start. Note that if any watch is slightly out of beat and has run down and stopped, the balance will probably not start automatically on winding but may require a flick of the wrist to start. As well as releasing the balance, the whip-balance stop gives it a gentle turning action so that it starts to vibrate immediately. It also releases the blocking lever which locks the seconds-recording wheel during the stop mode and return-to-zero mode.

Fig. 100 *The position of the levers when the watch is in the stop mode.*

Fig. 101 *Return-to-zero.*

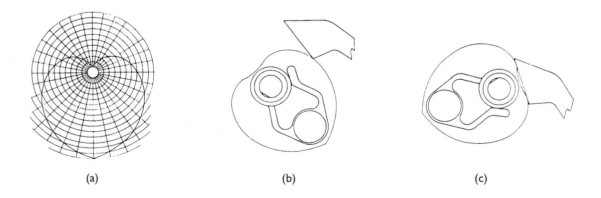

(a)   (b)   (c)

Fig. 102(a) *The design of the heart. (b) The seconds-recording heart being struck by the hammer. (c) Zero position.*

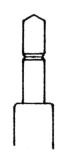

Fig. 103 The front pivot of the seconds-recording wheel with its groove.

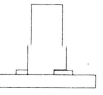

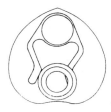

Fig. 104 The heart with its spring.

The seconds-recording wheel and the minute-recording wheel each have a groove for a spring on the heart to engage with (Fig. 103).

The heart has a hollow pipe with a groove: one leg of a spring passes into this, the other leg rests against the outside of the pipe (Fig. 104).

When the watch is reassembled, the grooves in the seconds- and the minute-recording wheels are lightly lubricated with a light oil. The hearts are then located in their respective grooves and held by the tension of the springs.

A fault which can arise is the heart slipping on its arbor, so that although a rate recorder may be showing a good rate, the hands indicate a loss of time. The reason for this is usually a worn spring in the heart. The combined effect of inertia due to the seconds recording, and starting and stopping five times a second and the worn spring, causes the heart to slip on the arbor instead of being carried by it. Invariably a new heart (or spring) cures the fault.

OVERHAULING A TYPICAL STOP WATCH (BFG 411)

Although the BFG 411 is now obsolete, there are still many in use which will need servicing, and it is a calibre that is representative of other calibres.

Dismantling

1. Wind the watch to check that the mainspring isn't faulty; operate the start, stop and return to zero to check that all three functions are working satisfactorily.
2. Remove the case back and check that the balance staff isn't broken.

When looking at the back of a stop watch, there is little evidence of the stop work mechanism, but you should see part of the whip-balance stop against the balance when the watch is in the stop or return-to-zero mode. (Fig. 105)

Fig. 105 The back of the watch with the stop work in stop and return-to-zero modes.

3. Remove the case screws and clamps. As parts are removed, they are put in the cleaning basket for cleaning.

4. Lift the movement out of the case by the button and pendant. With this particular calibre, the pendant is part of the movement and not part of the case.

Because of the difficulty in refitting stop-watch hands, it is usually better not to remove them from their hearts. Instead, the watch is put into the start mode, the dial screws are removed, then the hearts are pulled off from the seconds- and minute-recording wheels, taking the dial with them. All of this is achieved by holding each hand in turn under its centre by the nails of the thumb and middle finger, and pulling off. A tool could be used, but invariably I use the fingernail technique.

Fig. 106 Removing the seconds-recording hand with the special tool.

5. Remove both hands, the dial and the hearts as described above. For those who prefer to remove the hands from the hearts, Fig. 106 shows how to use the special tool.

6. Let the power off the mainspring by winding slightly, disengaging the click with pegwood and letting the spring down under control.

7. Still with the watch in the start mode, remove the four screws holding the cover for the stop work, and carefully lift off the cover so that bits don't fly under the influence of their tensioning springs. On no account operate the pillar wheel or parts may well fly.

8. Remove each of the four springs for the whip-balance stop, the jumper, the seconds hammer and the minutes hammer, remembering which goes where. Lift each spring off its post first, then release from the lever that each is associated with. To identify which spring goes where, I usually draw large circles in my bench paper and place each spring in a labelled circle. Surprisingly perhaps, the springs do not roll away.

9. Remove both hammers and the whip-balance stop.

10. Unscrew the pillar-wheel jumper and remove.

11. Unscrew the left-handed pillar-wheel screw and remove the pillar wheel, the steel core for the pillar wheel and the steel washer under the pillar wheel.

12. Remove the button and stem by first slackening the screw in the pendant. (Fig. 107)

13. Remove the spring for the operating lever.

Personally I would not remove the blocking lever as there is nothing to be gained by doing so, and it does take time to remove and replace. Leaving it in is one of the legitimate short-cuts that can be taken by repairers.

14. Remove both top and bottom shockproof units for the balance.

15. Remove the balance-cock screw, the balance cock and the balance from the movement.

16. Turn the cock over as you place it on the bench, turn the boot to release the balance

Fig. 107 The screw in the pendant for securing the stem.

spring, and either push the stud out of the stud holder if it is the movable type, or unpin the balance spring.

17. The balance may be cleaned in the cleaning machine in a mini basket, or in a spirit jar.
18. Remove the pallet-cock screw, cock and pallet.
19. Remove the ratchet-wheel screw (right-handed thread), click and click spring.
20. Remove the crown-wheel screw (left-handed thread), and the crown wheel. On this calibre there is no crown-wheel ring: instead, the crown-wheel screw is shouldered to form a bearing.

21. Remove the four identical screws from the train bridge and barrel bridge. (The previously removed case screws double as train and barrel-bridge screws.)
22. Remove the train and barrel bridges.
23. Remove the second wheel (minute-recording wheel), third wheel, escape wheel, barrel and fourth wheel (seconds-recording wheel).
24. Remove the clutch wheel and winding pinion.
25. Remove the screw holding the operating lever and the operating lever itself.
26. Remove the two screws holding the pendant and bow.
27. Remove the one remaining pivoted lever, operated by the clutch wheel when the button is pushed, by first undoing the shouldered screw. ('Pivoted lever' is a name I have invented because the sheet of part names I have for this watch doesn't include either a name or a number for this part.)
28. Dismantle the barrel for cleaning.

Clean the watch in the usual way in a cleaning machine.

Reassembly

1. Inspect holes in the bottom plate, pegging them out as necessary.
2. Replace the 'pivoted lever' – the last lever to be removed – and its shouldered screw after lubricating the shoulder of the screw. Check that it moves freely.
3. Replace the pendant and secure with two screws.
4. Turn the plate over, and lightly grease any rub marks on the plate caused by the operating lever.
5. Grease the post on the pivoted lever and replace the operating lever, tucking one end under the plate and dropping the other end over the greased post of the pivoted lever. Secure with its screw. Check that the lever is free.
6. Turn the watch over again and secure the plate in a movement holder.

7. Check the pivots of the seconds-recording wheel for wear, the pinion for cuts and cleanliness, and the teeth to make sure they are clean. It is most unusual to find bent or worn teeth here. Replace the wheel in the plate.

8. Inspect the barrel teeth to see that they are not bent, and are clean. If the mainspring has been removed from the barrel, replace it. Grease the spring, grease the pivot of the barrel arbor that operates in the barrel, and replace the barrel arbor. Grease the pivot that operates in the cover.

9. Replace the barrel cover. Align any marks on the outside of the barrel wall with the hole in the cover. Check the endshake of the arbor in the barrel: there must be endshake.

10. Grease the barrel-arbor pivots that operate in the plate and barrel bridge, and replace the barrel in the plate.

11. Check and replace the minute-recording wheel.

12. Peg the holes in the barrel bridge and replace, locating the minute-recording wheel pivot and securing it with its two screws.

13. Inspect and replace the third and escape wheels.

14. Peg the holes in the train bridge and replace carefully, locating the three pivots into their holes. Secure with two screws.

15. Check to see that all wheels can be lifted, and will drop down again under their own weight. Try this with the bottom plate facing up, and again facing down.

16. Replace the crown wheel. Grease the shoulder of the crown-wheel screw, and secure.

17. Check the freedom of the crown wheel.

18. Grease the pivoting point for the click, replace the click, replace the click spring. Grease the point where the spring presses against the click.

19. Replace the ratchet wheel and secure with the ratchet-wheel screw.

20. Using a screwdriver in the ratchet-wheel screw, wind the watch by about half a turn of the ratchet wheel, hold the ratchet wheel lightly against the click, and observe the train run down and reverse. This is a good sign that the train is free.

21. Inspect and replace the pallets, the pallet cock and pallet-cock screw.

22. Again with a screwdriver, wind the ratchet wheel by half a turn.

23. Jewelled pallet pivots are not oiled, but if the pallet pivots are not jewelled – that is, they work in brass holes – oil with Moebius 9010.

24. Put a drop of Moebius 941 oil on the impulse face of alternate scape teeth so that the oil will spread to all teeth and both pallet pins.

25. Oil the scape, the seconds-recording and the third wheel top pivots with Moebius 9010, and the minute-recording wheel top pivot with 9020.

26. After inspection, reassemble the balance and balance cock and put them back into the watch. Remember, jewelled impulse pins are not oiled.

27. Check the endshake of the balance, and that the balance spring is flat. Ensure that there is a small gap between the index pin and boot equal to about one-and-a-half times the thickness of the balance spring, and that when the balance is in the position of rest, the spring is touching neither index pin nor boot.

28. Check that the balance is in beat (see p. 45).

29. Turn the watch over and oil the bottom pivots in a similar way to the top (that is, the minute-recording wheel with Moebius 9020, and the third, the seconds-recording and the escape wheels with Moebius 9010. (Oil the bottom pallet pivot with Moebius 9010 only if the pivot operates in a brass hole.)

30. Push the operating lever down, and after greasing the teeth in the clutch wheel, replace it.

31. Grease the teeth in the winding pinion that engage with the clutch wheel, and replace.

32. Grease the stem and replace. Secure the stem by tightening the screw in the pendant.

33. Grease the shoulder of the clutch wheel that engages with the 'pivoted lever'.

34. Replace the operating-lever spring, and grease the point where the spring bears against the lever.

35. Replace the steel washer under the pillar wheel. Make sure it is tucked under the nib (the operating end) of the operating lever.

36. Replace the pillar wheel core.

37. Grease the outside of the core.

38. Grease the underside of the pillar wheel, all fifteen teeth and the outside face of the pillars.

39. Replace the pillar wheel, making sure that the operating lever lies between two teeth.

40. Secure the left-hand thread pillar-wheel screw.

41. Hold the operating lever away from the pillar wheel, and check that it is free.

42. Grease the post for the jumper, and replace the jumper securing it with its own screw. Check that the jumper is free.

43. Locate one end of the jumper spring over the jumper and the other end over its post.

44. Press the button a few times to see the pillar wheel advance.

45. Check that the jumper holds the pillar wheel correctly. (Fig. 108.)

46. While holding the operating lever out of engagement with the pillar wheel, rock the pillar wheel slightly clockwise and then anti-clockwise. Each time it is released, the jumper must pull the pillar wheel back again. If this doesn't happen, first try

Fig. 108 The pillar wheel and jumper correctly located.

removing any roughness on the two acting faces of the jumper with an Arkansas stone, and burnishing. If the problem persists, try burnishing the fifteen teeth of the pillar wheel. Two other problems could be a weak jumper spring or lack of lubrication. Only after trying these alternatives would one replace either the pillar wheel or the jumper, or both.

47. Grease the post for the whip-balance stop, and replace. If necessary, turn the pillar wheel to allow the end of the whip balance to pass between two pillars.

48. Replace the whip-balance-stop spring.

49. Grease the pressure point for the blocking lever.

50. Grease the post for the seconds hammer, and replace.

51. Grease the post on the minutes hammer, and replace.

52. Holding the whip-balance stop down, turn the pillar wheel so that the watch is in the return-to-zero mode.

53. Replace the spring for the seconds hammer and the minute hammer.

54. Grease the point of contact between minute and seconds hammer.

55. Replace the stop-work cover plate.

56. Grease the pillars in the pillar wheel where engagement takes place between pillars and levers. (The outside faces have been greased.)

57. Test the stop, start and return to zero. Leave the watch in the stop mode.

58. If the hands have been left on the hearts, give the hearts a dry clean by wiping over with pith dipped in alcohol or with Rodico. (Replacing the hands on their respective hearts is covered on p. 90.)

59. Lightly oil the notch in the seconds- and minute-recording wheels with Moebius 9010.

60. Lightly grease the periphery of both hearts by dipping pegwood in grease, allowing this to soak into the wood, then passing the greased wood across the periphery of both hearts.

61. Replace the dial by first locating each heart on its respective arbor, then 'feel' the spring of the heart into its groove. If the hands are alternately lifted and pushed down, taking the hearts with them, it should be possible to feel when the springs are located in their respective grooves.

62. Replace both dial screws.

63. Operate the start, stop and return to zero, particularly to make sure that the hands do return to zero. If the seconds-recording hand doesn't return exactly to zero and is less than ⅕ second out, correct by bending the hand. The slight bend will not be evident.

If the hands have been removed, after step 57, lubricate the hearts and notches, as in steps 59 and 60, then proceed as follows:

(a) Replace the hearts on their arbors. The longer pipe heart is the seconds-recording heart.

(b) Replace the dial and dial screws.

(c) Put the watch in return-to-zero mode.

(d) Place the minute-recording hand in its pipe, and hold its tip exactly on zero while pushing the hand on securely. Do this in a series of pushes, and keep checking that the hand is not slewing around away from zero. If it is, try to pull it back with a twisting action as it is pushed further on; this I do with the back of my tweezers. (Fig. 109) Check that the spring in the heart is properly located in the groove in the seconds-recording wheel again.

(e) Now replace the seconds-recording hand in a way similar to that which was used for replacing the minutes-recording hand. Rather than push the hand on with the back of the tweezers, use a hollow pusher made from an unwanted plastic knitting needle as it is less likely to leave a mark. Provided the hand is within ⅕ second, it may be bent to zero exactly as mentioned earlier. Check that the spring in the heart is properly located in its groove.

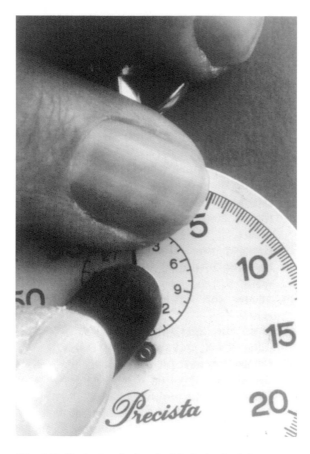

Fig. 109 *Replacing the hand with the back of the tweezers.*

64. Having cleaned the case, replace the movement and secure with case clamps and screws.

65. Replace the case back.

There are occasions when one or more levers in a stop watch will need to be modified. For example, when using a stop watch, operators often take up the play in the button and stem before starting, stopping and returning to zero. When the watch is in the start mode, taking up the slack can sometimes stop the watch before the pillar wheel is properly advanced. This is because the pillar wheel starts to advance, and a pillar moves the tail of the whip-balance stop, which stops the watch. (Fig. 110)

Fig. 110 If a pillar moves the whip-balance stop prematurely, a little metal on the whip-balance stop may need to be stoned away and the end polished.

The cure may be to remove a little of the metal on the whip-balance stop so that the pillars cannot make contact until the tip of one of the teeth in the pillar wheel passes the point on the jumper. Strictly speaking, no action should take place on a stop watch until a tooth tip on the pillar wheel passes the point of the jumper and

the jumper itself takes over the advancement of the pillar wheel. Actions are initiated by the operator, but are completed by the jumper. With some stop watches this is not achievable, so don't take any chances by modifying if replacement parts are not available.

Another point to watch for with a stop watch is that when initiating the start mode, the seconds-recording hand remains at zero until the pillar wheel jumps. Sometimes the hand can be seen to move slightly as the slack in the button is taken up due to hammer pressure coming off, and a worn pivot hole.

Just what can, or should be done about the problem mentioned above will depend upon the quality and design of the watch and the purpose to which it is to be put. Over the years I have serviced stop watches used for athletics, for timing dynamite fuses and for timing vehicles over a distance, and these activities have all necessitated correctly functioning watches.

7 CHRONOGRAPH OVERHAUL

In addition to showing the time of day, a chronograph has facilities for short interval timing which work independently. Chronographs operate on one or two pushers offering 'start', 'stop' and 'return to zero'. Teeth in the coupling clutch wheel engage with alternate teeth in the chronograph runner (sometimes called the seconds-recording wheel) which are very fine indeed. This has the advantage of minimizing any jump in the seconds-recording hand as the two wheels are brought into engagement when the chronograph is started. To assist this, these two wheels and the chronograph driving wheel have triangular teeth.

Currently there are four types of chronograph: two mechanical and two quartz. Of the two mechanical, one is what I have come to think of as lever-operated, the other is operated by a pillar wheel. The two types of quartz chronograph are electro-mechanical or all electronic, where timing is often to $\frac{1}{100}$ second instead of to the more usual $\frac{1}{5}$ second.

THE LEVER-OPERATED CHRONOGRAPH

For this exercise we will be looking at the Valjoux 7733 lever-operated type that utilizes two pushers (Fig. 111): the one at the 2 o'clock position allows the start/stop function, whilst the other at 4 o'clock operates the return to zero.

Fig. 111 Valjoux 7733 two-pusher chronograph.

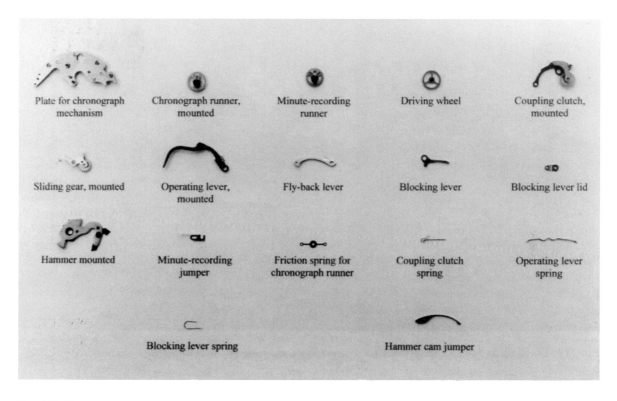

Fig. 112 Part names.

The terminology adopted here is that used by the manufacturer, and is likely to be the most helpful if spares are required. (Fig. 112)

Dismantling

1. Do the usual checks to establish why the watch needs attention.
2. Remove the two case screws and case clamps.
3. Push on the setting-lever post, withdraw the stem and button, and lift the movement from the case.
4. Replace the stem and button. The stem should simply push back in, though it may help to push on the setting lever again.
5. Ensure that the hands for recording hours, minutes and seconds are in a suitable position for removal, cover them and the dial with tissue paper, and remove all three hands together with hand-lifters. Sometimes the seconds-recording hand pulls off its collet, but this is not a serious problem as the collet can be pulled off after the cannon pinion has been removed, and the hand riveted back onto the collet. (It happened on my demonstration watch.)
6. Place tissue paper over the dial again and remove the small seconds hand and the minute-recording hand. These are not always interchangeable, so identify which is which.
7. Slacken the two side dial screws and lift off the dial. Tack down the side dial screws again to prevent them coming out in the cleaning machine.
8. Holding the watch in a movement holder, remove the power from the watch. The best access to the click is achieved with the chronograph in the return-to-zero mode. The click can be seen through the hole in the

Fig. 113 The click can be seen close to where the hammer cam jumper engages with the hammer mounted.

plate close to where the hammer cam jumper engages with the hammer mounted. You will have to turn the button slightly to see the click. (Fig. 113)

As work progresses, load the cleaning basket, putting the more vulnerable parts such as the chronograph runner into a mini basket. To save mixing up screws, some could be replaced in their respective holes for cleaning. It will save time in the long run.

9. Remove the Incabloc unit in the balance cock.

10. Remove the balance-cock screw and the balance cock.

11. With the balance cock upside down and flat on the bench, turn the boot by 90° to release the spring, then lift the cock and slacken the stud screw sufficiently to remove the stud. After removal, tack down the stud screw to prevent it coming out in the cleaning basket.

12. Check that the chronograph is in the 'go' mode: the hammers should be raised back, and the coupling clutch wheel should be engaged with the chronograph runner. Also the blocking lever (sometimes called the brake) should be clear of the chronograph runner.

13. Remove the screw, holding the minute-recording jumper, and then remove the minute-recording jumper itself. (In older chronographs this jumper is one of the most vulnerable parts of the watch and is often irreplaceable. Take extreme measures with older chronographs to prevent damage to the minute-recording jumper.) Remove the large screw only. The other one is, in fact, not a screw at all but an eccentric adjusting plug, and at this stage it is reasonable to assume that adjusting plugs are correctly adjusted; therefore time can be saved by leaving them alone. Adjustments will be checked later and any necessary alterations made.

Minute-recording jumpers should not be put in the cleaning machine but should be cleaned by hand in a spirit jar, and dried with pith or absorbent paper.

14. Remove the cock holding down the minute-recording runner and the chronograph runner. (Double check that the chrono is in the 'go' mode first.)
15. Carefully remove the chronograph runner, holding with tweezers on the crossings – never on the teeth. The teeth are very fine, and we need to prevent burrs from being flattened into the tooth spaces. (The friction spring for the chronograph runner may be removed now or later.)
16. Remove the screw for the hammer mounted. Holding the tension off the hammer cam jumper, remove the hammer mounted.
17. Remove the hammer cam jumper after removing its screw.
18. Remove the one large shouldered screw in the coupling clutch mounted, and remove the coupling clutch mounted. Again you will see two associated eccentric adjusting plugs that look like screws, but that, again, should be left.
19. Remove the coupling clutch spring and its screw to gain access later to the escape and fourth wheel cock screw.
20. Remove the chronograph driving wheel from the fourth wheel by feeding suitable lifters under the wheel or between the crossings, lifting under the boss. Note which way up the wheel is (the large boss facing down). (Fig. 114)
21. Remove the friction spring for the chronograph runner, if it was not removed earlier.
22. Take the blocking-lever spring on the sliding gear mounted to the other side of the post in the blocking lever, and remove the spring.
23. Remove the screw holding the blocking-lever lid, which in turn holds down the blocking lever; remove the lid and the blocking lever.
24. Remove the minute-recording runner.

Fig. 114 Using hand-lifters to remove the chronograph driving wheel.

25. Remove the shouldered screw holding the sliding gear mounted.
26. Remove the sliding gear mounted.
27. Remove the screw for the operating lever mounted, and the lever.
28. Remove the screw for the fly-back lever, and the lever.
29. Remove the two screws holding the chronograph mechanism plate, and the plate itself.
30. Remove the operating-lever spring.

What is left is an ordinary mechanism, except that the crown-wheel core has a hole for the minutes-recording runner arbor to pass through. Remember to replace it in the right position at the reassembly stage. On some chronographs, the ratchet wheel and crown wheel are positioned under the train bridge, and if the chronograph work is replaced with the crown-wheel core 180° out, the chronograph will have to be stripped again to correct this error.

Stripping and reassembling a chronograph is quite involved, and it is for this reason that it was usual to replace a steel mainspring automatically at the time of overhaul. It is not necessary to replace an unbreakable mainspring as a matter of course – though inevitably some repairers will prefer to do so.

The combined train and barrel bridge have two eccentric adjusting plugs which should be left, and there is an adjusting plug in the combined fourth and scape cock.

Clean the watch in the usual way, and inspect, reassemble and lubricate the basic movement up to the point where you would replace the pallets. The pallets and balance are not put in yet, as there are some tests associated with the chronograph work that have to be carried out first. Lubrication of the upper train wheels is done now, as some pivots will be inaccessible after the chrono work is put on.

Reassembly of the Chronograph Work

With the exception of lubrication, reassembly is much the same process in reverse. To save repetition, remember to check the freedom of each part after tightening its screw: this is done before loading any springs. If a component that is supposed to be free is pinched after its shouldered screw is tightened, it may be necessary to put the screw in the lathe and turn back the underside of the head slightly.

1. Replace the operating-lever spring, locating the ends of the spring in the cutouts in the plate.
2. Replace the chronograph mechanism plate and its two screws.
3. Grease the post for the fly-back lever and replace it and its screw. Make sure it is properly located with the operating-lever spring, and that the pin in the free end of the lever locates in the circular recess in the plate.
4. Grease the post for the operating lever mounted, and replace this, together with its screw. Remember to grease the spring which is a fixture on the operating lever, and the pivoting point of the lever it rests against.
5. Grease the pivoting point of the sliding gear mounted, and replace this with its shouldered screw. The pivots of the wheel are not lubricated.
6. Replace the minutes-recording runner (the pivots are not lubricated); engage the teeth with those of the wheel in the sliding gear mounted.
7. Grease the pivoting point for the blocking lever, and replace the blocking lever, the blocking-lever lid and its screw.
8. Replace and load the blocking-lever spring. Grease the point where the spring bears against the pin in the blocking lever.
9. Replace the chrono driving wheel onto the fourth wheel top pivot. Place the wheel over the pivot with the large boss facing down, then push the wheel on with the back of the tweezers, holding these parallel with the plate. Remember you will be pushing against a jewel. Great care must be taken to keep the pivot straight. Special tools are desirable to push this wheel on while supporting the bottom pivot of the fourth wheel, but with care, the job can be done successfully as described.

10. Wind the watch by a couple of turns of the button to confirm that the driving wheel is turning true. (Some watchmakers prefer to assemble and rate the basic movement before adding the chrono work in order to positively differentiate between any problems with the basic train and the chrono work. If this is done, just remove the balance and pallets for the above test and replace them again.)

11. Replace the pallets, wind the watch slightly, lubricate the pallet stones but not the pivots, check the escapement functions as with any watch, and replace the balance followed by the top shock system. Complete all the usual checks including endshake, balance spring adjustment and beat setting.

12. Replace the friction spring for the chronograph runner. It should be perfectly flat.

13. Replace the coupling clutch spring and its screw. This screw is not shouldered, and it is easy to carry the spring around as the screw is tightened: prevent this by holding the spring steady with a piece of pegwood. (Alternatively slacken and retighten the screw after replacing the coupling clutch mounted.)

14. Grease the pivoting point of the coupling clutch mounted, and oil the pivots of the wheel with Moebius 9010; then replace, checking that the teeth in both wheels engage safely. Secure with the shouldered screw, remembering to put a little grease under its head. Hold the tension off the spring to check the freedom of the coupling clutch mounted.

Check the height of the chronograph driving wheel to ensure that it is at least one-third engaged when both endshakes are opposed. Full engagement in the horizontal plane is ideal. If necessary, the chronograph driving wheel can be lowered by pushing with a flat hollow punch. (Fig. 115)

Now check the depth of engagement between the two wheels when looking from the top of the wheels. Engagement should be deep, but if the chronograph driving wheel is held stationary, it must be possible to rock the wheel in the coupling clutch. Check four points around the wheel at 90° apart: this will cover the instance of an out-of-round wheel and bent pivot. (Fig. 116)

15. Replace the hammer cam jumper.
16. Grease the post for the hammer mounted, and replace it as if the chronograph were in

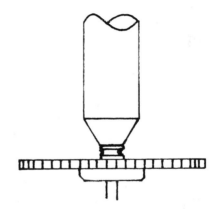

Fig. 115 Pushing the chronograph driving wheel onto the fourth wheel pivot with a flat hollow plastic pusher made from a knitting needle. If possible, support the bottom pivot using a staking outfit.

Fig. 116 Showing freedom between the chrono driving wheel and the wheel in the coupling clutch. It is adjusted by the eccentric plug which is the pivoting point of the coupling clutch.

the start mode. Hold the hammer cam jumper away from the hammer cam as the hammer mounted is replaced, then check the location of the spring and the sliding gear mounted. Replace the screw.

17. Hold the tension off the spring while you feel the freedom of the hammer mounted.

18. Grease the hammer cam jumper where it engages with the hammer mounted.

19. Confirm that the friction spring for the chronograph runner has not moved, then, with Moebius 9010, lightly oil the shoulder of the chronograph runner which bears against the spring, and replace it in the watch. Remember not to hold the wheel by its teeth.

20. Replace the cock for the minute-recording runner and the chronograph runner, taking care to protect the pivots and the teeth of the wheels. This must be done with the chronograph in the start mode.

21. Check the endshakes of both wheels.

22. Lubricate with Moebius 9010 the top pivot of the chronograph runner, but not the top pivot of the minute-recording runner.

23. Check the depth of the wheel in the coupling clutch mounted and the chronograph runner; a strong eyeglass will be needed for this. (Some watchmakers use a microscope to make this check, though it is not essential.) (Fig. 117)

Fig. 117 Check the depth between the coupling clutch wheel and the chronograph runner.

24. Very lightly lubricate the hammer faces with Microtime grease and operate the stop, the return to zero and the start a few times.

On chronographs where seconds and minutes hammers are rigid and combined, check that after a return to zero the chronograph runner is held securely at zero. Very slight movement of the minute-recording runner is permissible provided the minute-recording jumper holds the hand at zero immediately the chronograph is started. If, on return to zero, the minute-recording runner is held securely but there is some play between the chronograph runner and the hammer, stone a little off the front face of the minutes hammer with an Arkansas stone, and burnish. Ideally both will be held securely, though some slight play between the minute-recording runner and the hammer is acceptable.

25. Double check the depthing between the wheel in the coupling clutch and the chronograph runner.

26. Make sure the watch is in the return-to-zero mode. Replace the minute-recording jumper so that the end of the spring which engages with the teeth of the minute-recording runner is centralized between two teeth.

27. Put the watch in start mode, and check the engagement between the finger in the chronograph runner and the wheel in the sliding gear mounted. Check also that the spring isn't too strong, causing the watch to stop as the finger in the chronograph runner advances the wheel in the sliding gear mounted. For this test, you only want the mainspring to be wound by about one turn of the button. Spring tension can be adjusted by rotating the minute-recording jumper about its fixing screw after slackening the screw slightly.

28. Check that before the finger fully advances a tooth in the sliding gear, the minutes-jumper advances a tooth in the minute-recording runner. With this arrangement the minute-recording hand is stationary for most of the

time, advancing immediately after a complete minute is up – the hand is seen to jump as the tip of the tooth passes the point of the minute-recording jumper.

29. Grease the end of the operating lever where it engages with the hammer mounted, where the pin in the fly-back lever engages the hammer mounted, and the end of the sliding gear mounted where it engages with the hammer mounted.

30. When reassembling the motion work, remember to oil the front pivot of the chronograph runner, but not the minute-recording runner front pivot: like the back pivot, it runs dry.

31. Replace the dial and the hour, minute and small seconds hands in the usual way.

32. Make sure the chronograph is in the return-to-zero mode, and replace the minutes- and seconds-recording hands. They must be firm, and must zero exactly.

Terminology can vary slightly for chronograph parts, but the following is generally true for all chronographs and should be adopted in the absence of specific instructions by the manufacturer:

Fig. 118 A Lemania single pusher chonograph with pillar wheel.

- All points of heavy friction for chronograph operation should be lightly greased, including pivoting points and where springs exert pressure.
- Seconds- and minutes-hammer faces must be very lightly greased.
- Oil the upper and the lower pivots of the chronograph runner.
- Oil the upper and the lower pivots of the coupling clutch wheel.
- Oil the friction spring for the chronograph runner.
- Never oil the pivots of the minutes-recording runner.
- Never oil the pivots of the sliding gear wheel.

THE CHRONOGRAPH WITH PILLAR WHEEL

For this exercise we shall be dealing with a Lemania chronograph (Figs. 118, 119). The movement has strong similarities to the lever-operated chronograph already described, except that it has a single pusher offering start, stop and return to zero operated by a pillar wheel. There are six pillars operating four levers: each lever is held by a pillar for two pushes of the single pusher, whilst a third push allows each lever to move between the pillars for chronograph operation.

The minute-recording runner is advanced by an oscillating pinion driven constantly by the centre wheel. The oscillating pinion has a normal steel pinion at one end driven by the centre wheel, while at the other end is a pinion which engages with the minute-recording runner mounted during the start mode. To accommodate this, the pinion rocks on the bottom steel pivot. The minute-recording wheel is advanced continuously rather than in one-minute jumps.

Description of the Modes
It is important to understand what should be happening within the watch during a complete cycle of the three modes: return to zero, start

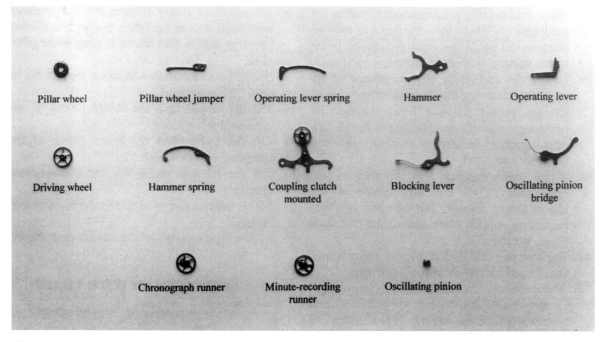

Fig. 119 Part names of the Lemania calibre 2210.

and stop. This is more easily done if you have a chronograph in front of you to look at as you follow the text through.

Return to Zero Mode

The tail of the hammer is between two pillars, the hammers themselves having returned to zero, the chronograph runner mounted and the minute-recording runner mounted. The coupling clutch mounted is held away from the chronograph runner mounted, and the blocking lever is off the chronograph runner. The lever holding the oscillating pinion top pivot keeps the other pinion away from the minute-recording runner mounted.

Start Mode

At the first push of the pusher, the hammers come off the hearts, the wheel in the coupling clutch engages with the chronograph runner mounted, and the oscillating pinion engages with the minute-recording runner. The blocking lever remains off. In this mode, the minute-

recording hand advances proportionally with the seconds-recording hand.

Stop Mode

At the second push the coupling clutch mounted disengages with the chronograph runner mounted, and the oscillating pinion disengages with the minute-recording runner mounted. The blocking lever engages the chronograph runner mounted, but the hammers remain off the hearts.

Return to Zero

At the third push, the blocking lever disengages with the chronograph runner mounted, and the hammers then fall, striking the hearts for the return to zero. The coupling clutch mounted and the oscillating pinion remain disengaged.

Dismantling

1. De-case, remove the hands, dial and hour wheel, and replace the stem.

2. Remove the power by turning the button slightly, releasing the click and letting the spring down under control.

3. Make sure the chronograph is in the 'go' mode.

4. Remove the balance-cock screw and the cock.

5. Remove the balance from the cock, the endpiece screws, the endpiece and the regulator.

6. Remove the pallet-cock screw, the pallet cock and the pallets.

7. Remove the cock screw and the cock of the minute-recording runner mounted and the chronograph runner mounted, and the runners themselves. Remember to hold the wheel crossings, and not the teeth.

8. Slacken the shouldered bridge screw for the oscillating pinion, and unlock the spring tail – which is part of the bridge – by passing it to the other side of the post it rests against. Remove the screw, bridge and the oscillating pinion. Hold the pinion carefully by its pivot – *not* by the pinion, for fear of damaging its fine teeth. Once removed, however, it is quite safe to hold it by the bottom pinion which is made of steel and quite robust.

9. Slacken the shouldered blocking-lever screw, lift the end of the spring over its post, and remove the blocking lever.

10. Remove the large- and small-shouldered screw holding the coupling clutch mounted, and remove the clutch also; it is not necessary to remove the latter's wheel. Do not turn any of the eccentrics at this stage – in fact it is better on principle to leave eccentrics until testing the chronograph work, and even then only to adjust them if it is necessary.

11. Slacken the hammer-spring screw, lift the end of the hammer spring to rest on top of the hammer, then remove the screw and the hammer spring.

12. Remove the shouldered screw holding the hammer, and the hammer itself.

13. Remove the driving wheel (the large boss faces up).

14. Slacken the screw holding the operating-lever spring, and take the tension off the spring by passing the end of the spring over the screw in the operating lever. Remove the screw and the spring.

15. Remove the operating lever and its screw.

16. Remove the pillar-wheel jumper screw.

17. Remove the pillar-wheel jumper spring by unlocking it first, then lifting it out.

18. Remove the left-handed pillar-wheel screw, and then the pillar wheel. You are now left with the friction springs for the chronograph runner and the minute-recording runner. These may be left on the top plate for cleaning. (Fig. 120)

Reassembly

When reassembling the basic watch, remember to secure the crown-wheel core so that the arbor of the minutes-recording runner can pass through. Oil the basic watch in the usual way, but if the pallet and balance have been replaced, remove them.

1. Grease the post for the pillar wheel.

2. Grease the teeth of the pillar wheel and the pillars, then replace the pillar wheel.

3. Grease the top of the pillar wheel that will contact the pillar-wheel screw, and replace the screw. Check that the pillar wheel turns freely.

4. Replace the pillar-wheel jumper, initially allowing the end of the jumper to rest on top of the ratchet-shape teeth in the pillar wheel.

5. Replace the screw, but don't tighten it yet: load the jumper spring first, then tighten the screw.

6. Rock the pillar wheel to make sure that the jumper pulls the pillar wheel back to its correct position from both directions.

7. Replace the operating lever, greasing the plate where there are any rub marks made by the operating lever.

8. Replace the screw holding the operating lever, and check that the operating lever is free. Put a little grease between the operating lever and the screw.

Fig. 120 The Lemania stripped of its chronograph work.

9. Replace the operating-lever spring. At first just tack down the screw, load the spring, then tighten the screw.

10. Put a little grease where the operating-lever spring bears against the small screw in the operating lever.

11. Test the action of the pillar wheel by pushing on the operating lever: the pillar wheel should advance smoothly by one tooth at a time.

12. Replace the chronograph driving wheel. Provided the pallet is left out, it should be possible to wind the watch by a couple of turns to check that the driving wheel turns truly, and that the fourth-wheel top pivot is straight.

13. Grease the post for the hammer, and replace the hammer and its shouldered securing screw. Check that the hammer is free, and grease where the hammer bears against the screw. The pillar wheel may have to be operated in the return-to-zero mode to test the freedom of the hammer for the full arc of operation.

14. Replace the hammer spring: leave its screw loose at first, then load the spring, then tighten the screw.

15. Grease the point where the hammer spring bears against the hammer.

16. Put the watch into the 'go' mode.

17. Grease the post for the coupling clutch mounted, oil the pivots of the wheel in the coupling clutch mounted, and replace it. Assist the teeth to engage with the driving wheel. Replace the large- and small-shouldered screws associated with the coupling clutch mounted. Check that it is free. Put a little grease under the two screws.

18. Check the depth of the driving wheel, and the wheel in the coupling clutch mounted. Any adjustments are made to the eccentric about which the coupling clutch mounted pivots.

19. Grease the shoulder of the blocking-lever screw, then replace the blocking lever, tighten the screw, then load the spring. Check that the blocking lever is free.

20. With Moebius 9020, oil the hole in the plate which accepts the oscillating pinion's bottom pivot. Replace the oscillating pinion with the steel pinion facing down, holding the top pivot and feeling the bottom pivot into its hole and seeing the pinion engage with the centre wheel.

21. Now for the tricky bit: replacing the oscillating pinion bridge. I find it best to locate the pivot first, then drop the bridge into position with the hole for the shouldered screw over the hole in the plate. Grease the shoulder of the screw, and replace it without the pivot coming out of its hole. To achieve this I hold the bridge down with pegwood. You only need to stop the bridge from rising up over the pivot.

22. Rock the bridge to check that it is free.

23. Load the spring by passing it to the correct side of the post in the coupling clutch, then grease between the post and spring.

24. Oil the top pivot of the oscillating pinion with Moebius 9020, and check its endshake.

25. Put a small drop of oil on the tension springs of the chronograph runner and the minute-recording runner where they engage with the underside of the hearts.

26. Check that the chronograph is in the 'go' mode. Lightly grease the periphery of both hearts with watch grease by allowing grease to soak into pegwood and passing the pegwood over the periphery of the hearts. Oil the front pivot of the chronograph runner with Moebius 9010, and replace the chronograph runner and the minute-recording runner. Remember that the pivots of the latter are not oiled. As you replace both runners, make sure the hearts are on top of the friction springs, not beside them.

27. Replace the bridge for holding the two runners. Take care to see that no teeth or pivots get damaged. My technique is to tighten each screw a little at a time, checking the endshake of both runners and seeing that the teeth in the chronograph runner engage safely with the teeth in the coupling clutch wheel, and that the teeth of the minutes-recording runner mesh safely with the oscillating pinion.

28. Check the depth between the chronograph runner and the coupling clutch wheel; also the depth between the minutes-recording runner and the oscillating pinion. Adjust if necessary. The eccentrics for each are quite obvious.

29. Now we need to check the clearance between the friction spring and the arbor of each runner. It is comparatively easy to check that between the arbor of the minutes-recording runner and its friction spring; however, between the chronograph runner and its friction spring is a little less easy to see. In fact it needs to be minimal, and I have come to rely on feel as much as sight. With a clean, fine, dry oiler, push on the side of the friction spring and see that it makes a small movement before it touches the arbor: the movement will be less than the distance of the closest point of the heart to the arbor, but provided it is positive, all will be well. (Fig. 121)

30. Oil the top pivot of the chronograph runner with Moebius 9010.

31. Wind the watch by a couple of teeth of the ratchet wheel to check that all is free.

32. Replace the pallet and balance, oiling and making adjustments in the usual way.

33. Operate the stop, return to zero and start a few times, then check the depth between the coupling clutch wheel and chronograph runner, and the oscillating pinion and the minutes-recording runner.

34. Replace the hour wheel, dial and hands.

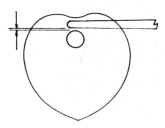

Fig. 121 Checking the clearance between the friction spring and the chronograph runner.

8 AUTOMATIC WATCHES

Automatic watches work on the pendulum principle: that is, if a weight is free to swing about an axis which does not pass through the centre of gravity of the weight, then when the axis of rotation approaches horizontal, the heaviest part of the weight will fall beneath it. In the automatic watch, the weight is known as an oscillating weight (sometimes called a rotor) and it turns by gravitational force through a part of a circle stopping against buffers, or a complete circle; it can wind in one or both directions. Winding through a complete circle in both directions is considered the best method, as the watch will be wound with less movement than one in which the oscillating weight has a limited arc and winds in one direction. This is particularly noticeable for people who are relatively inactive. The winding is effected through an automatic train which is usually mounted over the normal time train and drives the ratchet wheel through a series of reduction gears (Fig. 122). Because the watch winds automatically, the following contingencies should be considered:

Fig. 122 An automatic watch.

(a) Once the mainspring is fully wound, any further winding would cause damage, perhaps by breaking teeth out of wheels, so a clutch arrangement needs to be built in. This usually takes the form of a stiff, short length of spring, the same height as the mainspring, fastened to the outer end of the mainspring to act as a slipping clutch. Once the mainspring is wound, the stiffer spring to which the mainspring is secured slips around the inside wall of the barrel: obviously this must be suitably lubricated, and a special graphite lubricant is available (Moebius 8301).

(b) An additional arrangement is necessary as a means of holding the ratchet wheel without a whole tooth having to pass the click. This way, even slight movements of the oscillating weight contribute to the winding of the mainspring. It is achieved by a holding arrangement, sometimes a second click, in the automatic work. Without this it would be possible for the mainspring to turn the oscillating weight when the watch is in the horizontal position – which defeats the object.

(c) It is necessary to accommodate a two-directional oscillating weight driving a ratchet wheel that turns in one direction only.

(d) Often the crown wheel and winding pinion also turn during automatic winding, so consideration is given to reducing frictional loss here.

OVERHAULING AN AUTOMATIC WATCH

For this exercise I have chosen an ETA 2472. Much of what is said is transferable to other calibres, something we all have to do when technical information is not available. Some calibres of automatic watches are significantly different, and I strongly recommend obtaining and following makers' technical guides where possible.

The Principle of Operation

For the ETA 2472 the oscillating weight carries a wheel on the underneath side – called a 'bearing wheel for oscillating weight' – the teeth of which engage with a 'pawl winding wheel complete' (with pinion) and an 'additional pawl winding wheel' (without pinion). Between them, they convert the movement of a two-directional oscillating weight to a one-direction reduction gear, which in turn drives the driving gear for the ratchet wheel, and the ratchet wheel itself. (See Fig. 123)

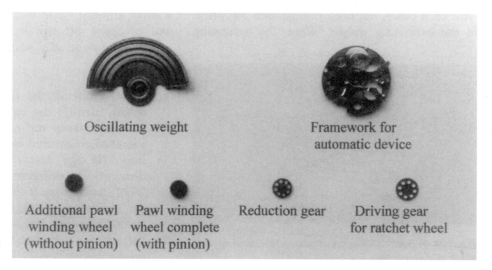

Oscillating weight

Framework for automatic device

Additional pawl winding wheel (without pinion)

Pawl winding wheel complete (with pinion)

Reduction gear

Driving gear for ratchet wheel

Fig. 123 Part names for automatic winding of calibre ETA 2472.

The two pawl winding wheels need to be checked at the time of overhaul. So that we are clear which side of the wheels we are looking at, let us call the side of each wheel that faces the back of the watch, the top. Hold the bottom wheel of the additional pawl winding wheel (without pinion) between your fingertips and rotate the upper wheel with your tweezers held in your other hand. With the bottom wheel held stationary, the top wheel should turn anticlockwise but lock up when turned clockwise. This is made possible because there are two pawls between the wheels which pass two sets of projections freely when the top wheel is turned in one direction, but which lock up when the wheel is turned in the opposite direction. One set of projections is on the inner part of the lower wheel, the other is on the outer part. (Fig. 124)

The other wheel is called the pawl winding wheel complete (with pinion). Hold the bottom wheel as before, and you should find that the upper wheel will turn clockwise only, locking up when turned in the opposite direction. This again is due to protrusions between the wheels and the pawls. If either of the two pawl winding wheels fails to lock up after cleaning, replace the faulty wheel; a repair is not practical.

In the watch, the two lower wheels turn in one direction only, and because they engage with each other, they must turn in opposite directions. The two upper wheels are not engaged with each other, but both engage with the wheel in the oscillating weight. When the oscillating weight turns clockwise, the upper and lower wheels of the pawl winding wheel complete (with pinion) turn anticlockwise because the pawls lock up on the protrusions in the lower wheel. At the same time, the upper wheel in the additional pawl winding wheel turns anticlockwise, but the lower wheel turns clockwise as the pawls pass the protrusions in the lower wheel freely.

When the oscillating weight turns anticlockwise, the upper wheel in the pawl winding wheel complete (with pinion) turns clockwise, while the lower wheel turns anticlockwise as the two pawls pass the protrusions freely. Both wheels of the additional pawl winding wheel turn clockwise as its two pawls lock up. The two pawl winding wheels also hold the ratchet wheel when it is wound by only part of a tooth. The click engages with the crown wheel instead of the more usual arrangement of engaging with the ratchet wheel. Check that the click is functioning properly before removing the automatic work.

Automatic watches are fairly trouble free, but look out for cut pinions in automatic wheels, and particularly for wear in the bearing for the oscillating weight. When the wear is bad, the oscillating weight will rub the watch plate and/or the case back; it is easily identified by worn plating on the watch or weight towards the outside of the movement. To test, hold the watch horizontally and lift the outside of the oscillating weight; a small amount of lift is necessary for freedom, but excessive lift will lead to the watch not winding automatically, and to worn plating.

Dismantling

1. After removing the hands, dial and hour wheel, identify the two screws holding the automatic work and remove them; they are sometimes coloured blue for easy identification. (If the watch you are about to dismantle is a different calibre from the ETA 2472, look to see if it is possible to remove the oscillating weight before removing the automatic work. If it is, remove it first.)
2. Lift off the automatic work complete with the oscillating weight and train. The ratchet

Fig. 124 A pawl winding wheel dismantled, showing two pawls and protrusions.

wheel will drop back by up to one tooth, but this does no harm even on a stopped watch with the mainspring virtually fully wound.

3. Place the automatic work on the bench with the underneath side facing up, and remove the screw holding the oscillating weight.

4. With the screw removed, lift the auto work off the oscillating weight and turn it over. The oscillating weight may be placed in the cleaning basket.

5. Turn the 'screw' retaining the pawl winding wheels to release one of the wheels, and again to release the other. In fact it is not a threaded screw, but is a friction fit in the plate, and has a boss with a flat for the wheels to pass. It will turn in either direction, and is not removed.

6. Check both pawl winding wheels and put them in the cleaning machine. They do not dismantle.

7. Turn the framework for automatic device over, and turn the bar holding the reduction gear and the driving gear for the ratchet wheel.

8. Remove the reduction gear and the driving gear for the ratchet wheel.

In this calibre, all of the automatic wheels rotate about posts secured to the framework for automatic device.

9. Remove the power from the basic watch, dismantle, and clean in the usual way.

It is recommended that the mainspring be removed from the barrel for cleaning both it, and the barrel, barrel cover and barrel arbor. An alternative may be to give the barrel a dry clean after removing the cover and arbor, then feed a little graphite grease between the inside wall of the barrel and the sipping end.

Reassembly

1. If the mainspring has been removed from the barrel, grease the inside wall of the barrel with special graphite grease, Moebius 8301, and then replace the mainspring using a watch mainspring winder. It is possible to replace it by hand, but extra care is necessary to avoid distorting the spring. Make sure your hands are dry and clean.

Start by replacing the outer coil, then feed the rest of the mainspring into the barrel, finishing of course with the inner coil. Most of the work is done with the index fingers and thumbs of both hands. Watch repairers will often grease previously used unbreakable mainsprings, though manufacturers claim they have life-long self-lubricating properties. Perhaps watch repairers feel uncomfortable with this after being used to greasing steel mainsprings.

2. Reassemble and lubricate the rest of the basic watch in the usual way.

3. Place the framework on the bench with the underneath side facing up. Lubricate the two posts in the framework for automatic device with 9020 and replace in turn the driving gear for the ratchet wheel (pinion up), and the reduction gear (pinion up).

4. Swing their retaining bar fully anticlockwise to lock them in place.

5. Lubricate the pivoting points of both pawls in both pawl winding wheels with 9020.

6. Turn the framework over and lubricate both posts for both pawl winding wheels with 9020, replace the pawl winding wheel complete (with pinion, facing down), and turn its retaining screw.

7. Replace the additional pawl winding wheel (without pinion) and turn the retaining screw again. Leave the flat of the screw facing the outside of the framework. (Fig. 125.)

Fig. 125 Leave the flat on the retaining screw facing the outside of the framework.

8. Lubricate with Moebius 9020 that part of the oscillating weight that bears against the framework for automatic device, and place the framework over the oscillating weight.

9. Lubricate with Moebius 9020 that part of the framework against which the screw for securing the oscillating weight bears, then replace the screw.

10. Check the freedom of the oscillating weight to rotate, then replace the whole automatic work on the watch. Usually it takes very little agitation of the oscillating weight for the driving gear to engage with the ratchet wheel.

11. Replace the retaining screws.

12. Check the freedom of the oscillating weight to rotate, also lift the outside edge of the oscillating weight to check the wear in the bearing. A lift of, say, 0.05mm should be in order, but the weight must never touch any other part of the watch or the case back. A further check is to rotate the watch about an axis passing through the three and nine o'clock positions. The oscillating weight should turn as the watch is rotated without tending to lock up.

When the watch is very nearly run down, the oscillating weight is likely to be very free but a little stiffer when the mainspring is so far wound that it slips around the wall of the barrel.

A fully wound automatic watch in good condition should run for about fifty hours without further winding.

9 FUSÉE WATCHES

Unlike contemporary mechanical watches which are invariably fitted with unbreakable mainsprings, early watches were fitted with steel mainsprings, hardened and tempered to dark blue. The comparatively poor timekeeping of early mechanical watches was due in part to the changing power output from the steel mainspring as the watch ran down. From fully wound there is an initial rather large disproportionate power drop-off, leading to a smoother drop-off in power, with a further disproportionate power drop towards the bottom end of the mainspring. (Fig. 126) This affects the amplitude of the balance which in turn, for a variety of reasons, leads to different times of vibration of the balance as the watch reaches different states of winding. Assuming there are no other influences on the balance, the watch would have a comparative loss as it runs down.

The fusée was incorporated in some early pocket watches to maintain a constant output of power even as the mainspring ran down, and was achieved through a toothless barrel driving a fusée wheel via a chain. Effectively, as the power of the mainspring diminishes, the chain exerts a turning force on the grooved fusée wheel on a gradually increasing diameter so that the turning force on the fusée wheel is much more constant. This of course leads to a more constant amplitude of the balance. (Fig. 127)

Frequently, maintaining power was a further refinement in fusée watches which kept power on the train while the watch was being wound. Maintaining power is recognizable by the presence of three components: a steel wheel with fine ratchet teeth between the fusée and wheel; a steel arbor with a type of pawl which engages with the steel wheel; and a steel spring keeping the pawl in engagement with the steel wheel.

The escapement may be a ratchet-tooth lever escapement or a verge. The significant difference with a verge escapement is that there is no pallet: instead, the balance staff has two 'flags' which act like pallets to receive impulse from a crown wheel (escape wheel). The 'action' is: impulse, drop, recoil; impulse, drop, recoil. The recoil is the slight reversal of the crown wheel during the supplementary arc of the balance.

DISMANTLING A FUSÉE POCKET WATCH

1. With a suitable pointed tool, push out the brass pin holding the bezel and the movement in position. Sometimes the bezel needs to be opened first to gain access to the pin holding just the movement. Invariably these push out from the 11 o'clock side to the 1 o'clock side. Remove the movement from the case.

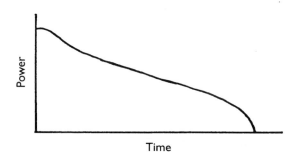

Fig. 126 A mainspring power curve.

Fig. 127 A fusée barrel without teeth, a fusée chain and a fusée wheel.

2. Remove the dust cover if fitted, and establish whether the escapement is lever type or verge.

   If it is a verge, there is an opportunity for letting off very nearly all the power by carefully holding the contrate wheel (the wheel before the escape wheel with teeth turned in a different plane to the rest of the train) and removing the balance. If the train is free, it should run down once the balance is removed. If the watch is a lever type, all the power, including the pre-load, will have to be let down with a key from the dial side.

3. Remove the balance cock holding the contrate wheel if a verge. Note the overhang of the balance spring beyond the stud. Still holding the contrate wheel, remove the pin holding the balance spring, feed the spring out of the stud and remove the balance itself. Allow the watch to run down. If a lever type, do the same except there will be no contrate wheel and the pallet will prevent the train from running.

4. Remove the hands.

5. Remove the three brass pins holding the dial backing plate, and remove the dial, the dial washer if fitted, the hour wheel and minute wheel.

You will notice the bottom pivot of the barrel is elongated and squared. On the square is a ratchet wheel, probably blued, with a click engaged with one of the teeth. Notice there is no click spring. This is because the screw holding the ratchet wheel does not have a shoulder and once the pre-load is put on the mainspring, the screw is secured which locks the click until the watch is serviced once again. Winding the watch is effected from the back of the watch on the fusée wheel, not the barrel. (Fig. 128)

6. Find a suitable key to fit the square protruding above the ratchet wheel. A pinchuck would do for this.

7. Hold the click in engagement with the ratchet wheel while you slacken the screw in the click by about half to one turn.

Fig. 128 Winding from the back of the watch.

suitable replacement if this happens. In principle the mainspring should be removed not only to clean it, but to assess its ability to provide adequate power to drive the watch. Steel springs are inclined to tire and to 'set', where the spring remains close-coiled when removed from the barrel.

I would argue that the purpose today of an old fusée pocket watch is not to keep very close time over a twenty-four-hour period, but rather to be perhaps a collector's item, a pleasure to own and wear occasionally, and to be seen as a modest investment. As good examples of fusée pocket watches become less common, conservation is becoming increasingly important.

8. Position the key over the square, push down so that it doesn't slip off the square, and turn it slightly to disengage the click; allow it to turn, letting the spring down, by about half a turn, then re-engage the click. Repeat this until all the power is off the mainspring. If a pinchuck is used instead of a key, it should be possible to let the power off by allowing the tool to slip under control between your fingers and thumb.

9. Lift the ratchet wheel off the squared barrel arbor bottom pivot.

10. Unscrew the two screws holding the barrel bridge, and remove the barrel bridge. Unhook the fusée chain from the barrel and remove the barrel from the watch. Be careful not to damage the end of the fusée chain.

11. Remove the barrel cover and barrel arbor, but, if it isn't damaged, leave the mainspring in the barrel for cleaning.

No doubt this recommendation will have its critics, but the alternative of removing and then replacing the mainspring will risk damaging it, and it will be difficult or impossible to find a

12. Remove the screws (or pins, some very short) holding the main top plate: if the watch has a lever escapement, lift the top plate, but only slightly at first so as to free the top pallet pivot from its hole; then remove the pallet, and only then lift the plate completely. This is because the end of the lever (the notch) lies between the top plate and the cock which supports the bottom balance pivot, and lifting the plate straight off will almost certainly break the bottom pallet pivot. There are no such problems if the escapement is a verge.

13. Remove the remaining wheels and the pawl for the maintaining power, and pull the cannon pinion from the centre wheel. There is no need to remove the maintaining-power pawl spring, the third and fourth wheel bridge, nor the stop-iron which is a lever to prevent the fusée chain coming off the end of the fusée wheel. When winding, eventually the fusée chain touches the stop-iron, lowering it so that the stop on the fusée wheel butts the free end of the stop-iron.

If the watch has a verge escapement, the crown wheel is likely to be held in the horizontal plane by a removable plug, in which case simply withdraw the plug and remove the wheel.

14. Remove the fusée chain from the fusée wheel, and clean in a spirit jar. After

Fig. 129 A fusée chain has two different ends.

cleaning, put a few drops of watch oil into a small quantity of clean degreasing agent, and drop the chain in for a few minutes. Remove the chain, and allow the degreasing agent on it to evaporate; this will leave the chain lightly lubricated. It should now be ready to put back in the watch. Note that the two ends of the fusée chain are different. (Fig. 129)

15. Remove any balance-cap jewels and jewel holes.
16. Dismantle the fusée wheel:
    (a) Push out the copper or brass pin in the steel collar holding the fusée wheel together (Fig. 130).
    (b) Remove the steel collar, the brass wheel with its maintaining-power spring, the

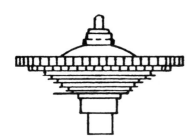

Fig. 130 The fusée wheel.

steel wheel for the maintaining power that the maintaining-power pawl engages with, and check the two clicks inside the steel wheel for wear: it is common to find a step worn in them, and for the tip of the clicks to be worn away completely. The various parts of a fusée wheel are shown in Fig. 131.

Making a New Click

Making a new click for the maintaining power wheel is not difficult, but it does help to have an overview of how to proceed. The first problem is how to hold the click as it is being made, and the simplest solution is just to make it with something to hold on to. A little judgement will be needed in deciding the length and diameter of the pivot of the click before riveting.

1. To drive out the old click, support the wheel over a hole in a stake and drive the rivet out with a pin punch. The clicks are quite soft, so the rivet should come away easily.
2. We are going to produce a 'blank' for the new click from carbon steel, leaving enough of the original material to hold on to until the last minute. Measure the diameter of the old rivet and asses the diameter necessary to form the new click. (Fig. 132)

Fig. 131 Parts of the fusée wheel.

Fig. 132 A blank for a click.

Fig. 133 The blank is filed to the outline.

3. Mark out the outline of the old click, and file to shape with a fine needle file (Fig. 133).

4. When very close to size, cut the pivot, leaving a piece long enough to use for holding, and try it on the wheel. Modify the shape as necessary until the click operates as the original would have done when new.

5. When you are satisfied, cut it to length and, having first stuck Sellotape to the wheel to protect it, file the cut end as close to the wheel as you can, finishing with a stone. This will throw out a burr almost sufficient to hold without riveting. Remove the Sellotape, which was there to protect the wheel from the file, and lightly rivet the new click, checking all the time that it isn't tight. The click must be left free, and the rivet must be flush with the wheel.

Reassembly

1. Have ready a copper taper pin filed to fit the fusée wheel arbor.

2. Inspect the top and bottom plate, remembering to peg any holes, whether brass or jewelled.

3. Check the operation of the stop-iron, and lubricate its pivoting point with 9010.

4. Check, lubricate and replace the balance jewel holes in the bottom plate and the balance cock.

5. Reassemble the fusée wheel, lubricating all friction points with 9020:

   (a) Lubricate the mating faces of the fusée and the maintaining-power wheel and the teeth of the ratchet wheel. Replace the steel maintaining-power wheel on the fusée, and engage the clicks by turning and lowering the wheel.

   (b) Lubricate the fusée arbor and replace the fusée wheel, lubricating the mating faces. Remember to locate the pin of the maintaining spring into the hole in the maintaining-power wheel.

   (c) Lubricate the point of contact between the collar and the wheel, and replace the collar. The hole may not be drilled centrally through the collar and arbor, so turn the collar if necessary to obtain a better alignment.

(d) Push the copper pin through the hole, and cut the larger side off flush with the hole, but leave a short overhang on the other side for gentle riveting. Cutting is achieved by pushing against the copper with a sharp screwdriver or knife, then bending the copper wire. It should break at the pressure point. After riveting, check that the arbor is free to turn. A slight resistance to turning is normal.

6. Grease and reassemble the barrel. Check the endshake of the arbor.

Because of the balance bottom cock and the danger to the pallet pivots, these watches, like Old English levers, are assembled into the top plate, then the bottom plate is replaced. Once the pivots are located, the movement is turned over to secure the plates.

7. After checking the pivots, teeth and leaves, replace in the top plate the fusée wheel, the centre wheel, the third wheel, fourth wheel and the pallet, remembering to put the notch under the balance bridge and the other end of the pallet between the banking pins. Finally, replace the escape wheel and the maintaining-power pawl.

8. Replace the bottom plate, locating the centre and fourth wheel pivots first. As mentioned above, once the pivots are located, turn the watch over and secure the plates.

9. The barrel bridge and the balance cock fit over two of the pillars. These pillars take very short pins – barely longer than the diameter of the pillar – and the cocks have countersinks to accommodate them. (Fig. 134)

10. Check the endshakes of all the wheels.

11. With a key, turn the fusée wheel so that the point for attaching the fusée chain is facing towards the outside of the watch.

12. Lubricate and replace the fusée barrel and barrel bridge and its two screws. Make sure

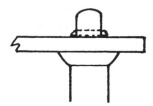

Fig. 134 The short pins are barely longer than the diameter of the pillar.

the bridge is in full contact with the train bridge and not resting on a pin.

13. From close to the barrel, feed the fusée chain behind the pillar and onto its anchoring point in the fusée wheel (Fig. 135).

14. Wind the chain onto the fusée wheel, making sure the chain doesn't jump a groove. You will have to guide the chain, keeping light tension on it by letting it run between your first and second finger.

Fig. 135 The fusée chain is attached to the fusée wheel.

Fig. 136 Keeping tension on the chain while winding it onto the fusée wheel.

15. Wind very nearly all the chain onto the fusée, just leaving enough to reach the barrel and hook the end into the hole in the barrel wall. Keep the chain taught while you replace the barrel ratchet wheel and engage the click just to take up the slack in the chain. (Fig. 136)

16. Next we are going to transfer the chain to the barrel. For this, remove the third and fourth wheel train bridge, and the third wheel so that the fusée wheel can turn.

17. Now turn the barrel arbor with a key on the square to transfer the chain to the barrel. As you do this, keep a finger on the fusée wheel or its square to create a slight drag; this ensures the chain comes off the fusée wheel smoothly. As you wind, hold the watch so that gravity causes the click to drop between successive ratchet wheel teeth. When all the chain has left the fusée wheel and is on the barrel, wind the ratchet wheel by just one tooth to keep tension on the chain – and to give yourself a well earned rest!

18. Replace the third wheel and the bridge.

19. Wind the barrel arbor by a half to one turn and, while holding the click into engagement with the ratchet wheel, lock down the click screw, leaving the click in full engagement with the ratchet wheel. This is the pre-load, which will not be altered unless the watch is stripped again. The pre-load could be adjusted up or down to make sure of a near-uniform power output from the mainspring throughout the going period of the watch, but it is hardly worth the effort. Provided the mainspring is the right length – that is, filling one third of the barrel space – no problems should be apparent by putting on up to one turn of pre-load. Personally I am happy with a slightly high pre-load as it can help to compensate for a tired mainspring that has been in the watch for over a hundred years. The mainspring must never be fully wound or there is the risk of damage to the watch. The limit of winding must be controlled by the stop-iron.

20. Oil the train.

21. Now wind the fusée chain back onto the fusée wheel, guiding it into the grooves in the wheel. This is only necessary for the first wind after overhaul. As the watch runs down, the chain will align itself naturally onto the barrel and will require no further guidance for winding again. As the watch becomes fully wound, check the operation of the stop-iron.

22. Oil both pallet stones with Moebius 941, and check the escapement functions.

23. Having washed the balance in degreasing agent and made any corrections to the balance spring, replace the balance in the watch, and feed the end of the spring into the stud, making sure that the spring is between the index pins; then pin the balance spring with its original brass pin.

24. Adjust the balance spring so that it will be flat, and will clear the plate and the arms of the balance.

25. Make sure the impulse pin is in the notch, and replace the balance cock.

26. Check the balance spring again, and make sure that when the balance is in the position of rest, the spring is central between the curb pins. Check the amplitude of the balance when a) the watch is fully wound, and b) when it is almost fully run down. If the amplitude drops off considerably as the mainspring is almost run down, the pre-load may need to be adjusted.

27. Lubricate the centre-wheel arbor to take the cannon pinion and the minute-wheel post, then replace the cannon pinion and check its tension.

28. Replace the minute wheel and the hour wheel.

29. Replace the dial and the hands.

30. Case the watch after cleaning the case.

10 BASIC ELECTRICITY AND MAGNETISM

It is quite possible to service quartz watches effectively without really understanding electricity itself, or the units used in measuring electrical quantities. The purpose of this book is, however, to dispel the mysteries surrounding watch repair, so the basics of electricity and magnetism have been explained, sufficient at least we hope for the reader to feel at ease servicing quartz watches. Certainly a better understanding of these will help you follow the working of a stepping motor (as shown on p. 128).

MAGNETISM AND ELECTROMAGNETISM

Magnetism can be said to exist where magnetic lines of force can be detected. We can detect a magnetic force by observing how a compass needle that is free to turn, always comes to point to the earth's magnetic north and south poles. What is actually happening is that the earth's magnetic north pole is attracting the opposite or south pole of the compass (sometimes called the north-seeking pole, or south pole) and that the earth's magnetic south pole attracts the north pole of the compass needle.

Some Facts about Magnetic Materials
1. All materials are magnetic to a greater or lesser extent but, in the present context, magnetic materials include iron and steel.
2. Magnetic materials are attracted to a magnet and, on contact, some will retain a degree of magnetism – for example, steel tweezers – whilst others will not. Thus, some antimagnetic tweezers are attracted to a magnet, but lose their magnetism when removed from its influence.
3. Magnetism can be induced into some materials, for example a steel needle, by stroking with a permanent magnet. One end of the needle will become a north pole, while the other will become a south pole. Which is the north and which the south will depend upon which end of the permanent magnet is used to stroke, and in which direction the stroking occurs. The hard steel needle itself now becomes a permanent magnet, yet the inducing magnet loses none of its strength.
4. Not all materials retain induced magnetism. For example, soft iron can be attracted to a permanent magnet and will itself attract other soft iron; however, once the permanent magnet is removed, magnetic forces in the soft iron will deteriorate and will no longer attract other soft iron. This can be demonstrated by a permanent magnet which will effectively hold a number of touching paper clips: if the permanent magnet (the inducing magnet) is removed, however, the paper clips are no longer mutually attractive.
5. Permanent magnets are made of hard steel and may be alloyed with other metals.
6. Every magnet has two poles, a north pole and a south pole.
7. Magnetic poles within a magnet are of equal strength.
8. The strongest part of a magnet is just back from its pole tips.
9. Like poles repel, unlike poles attract.
10. Magnetic lines of force can penetrate objects.

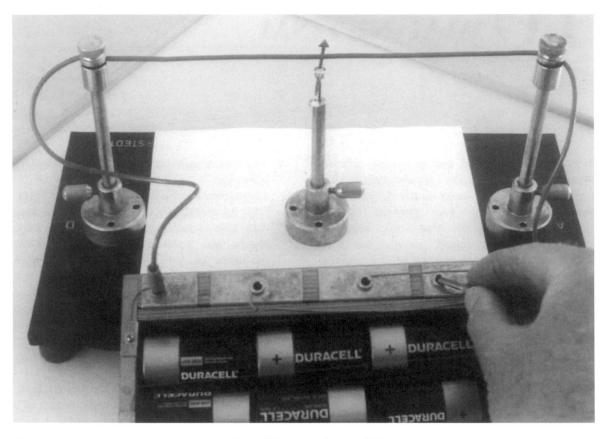

Fig. 137 Demonstrating that a compass needle parallel to a conductor will deflect by 90° when a current is passed through the conductor.

Electromagnets

If a current is passed through a conductor, a magnetic field will then surround it. This can be demonstrated by placing a compass needle which is pointing north/south parallel to a conductor while the conductor is connected momentarily to a battery: the needle will be seen to deflect, showing that there are lines of force present. (Fig. 137) Should you wish to try this for yourself, use insulated wire of about 1mm in diameter, connected briefly to a 1.5V battery. Be warned, however: it will drain the battery very quickly.

If the conductor is coiled, the field strength is increased, and it increases still further if the coil is wound around a piece of soft iron. This is because the soft iron itself becomes a magnet and contributes to the total field.

Just as we can produce magnetic lines of force by passing a current through a conductor, so a current can be produced by moving lines of force, cutting through a conductor. This is demonstrated with two coils connected to form a continuous loop, a permanent magnet and a galvanometer: as the magnet is lunged into and out of one coil, the galvanometer in the other coil deflects first one way, then the other. (Fig. 138)

A horological use for what we have been looking at is a stepping motor for a quartz watch (see p. 128), and a generator such as is used on the Seiko Kinetic (see p. 155).

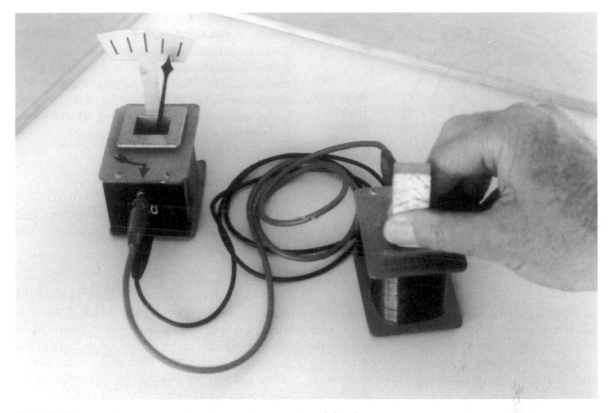

Fig. 138 Using a galvanometer to demonstrate the generation of electric current.

SIMPLE ATOMIC THEORY

At school we were taught the three states of matter: solid, liquid and gas. We were taught that most objects are compounds, and that if a compound object existed in the smallest quantity possible, it would be a molecule of the substance. A molecule of a compound is made up of elements, of which there are ninety-two known stable elements, and each element is made of atoms. As an example, water is a compound of two elements, hydrogen and oxygen, and a molecule of water (H_2O) contains two atoms of hydrogen and one atom of oxygen.

An atom consists of a central portion called the nucleus, which for most atoms is made up of a number of positively charged particles called protons, and a number of neutral particles called neutrons, which keep the protons in close proximity to each other without touching. Orbiting the nucleus are negatively charged particles called electrons, and where there are three or more electrons they are arranged in bands or shells; those in the outside band are called valence electrons. Under normal conditions, the number of electrons and protons is equal, and the charge between the two is balanced.

To exemplify this we will take the metal beryllium, which is sometimes used for making balance springs; it has four electrons, four protons and five neutrons (*see* Fig. 139). The beryllium atom's mass number is nine, which relates to the sum of the protons and neutrons, and it has an atomic number of four, which is the number of protons or electrons. Protons are about 1,800 times the mass of electrons, and if drawn to scale with a nucleus of 1cm diameter,

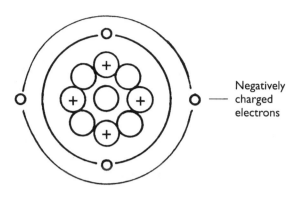

Negatively charged electrons

Fig. 139 Concept of a beryllium atom.

an electron would have to be drawn about half a kilometre away: from this it may be seen that matter is largely empty space. If it were possible to have a 1cm cube of densely packed protons and neutrons, the result would have a mass of three million tons!

As a point of interest, all electrons are the same, as are protons and neutrons, the difference between elements being the number of each.

Some elements have valence electrons which are not tightly bound to any one atom, but are capable of random movement between one atom and another. Electricity is the result of causing electron movement in one direction only: it can be exemplified by electrons moving in one direction through a copper wire. This wire is referred to as a conductor, and the electron movement in one direction is referred to as electric current. When electric current flows in one direction only it is referred to as 'direct current' or 'DC': when the current flow reverses in a regular manner, it is referred to as 'alternating current' or AC.

There are elements and compounds that strongly resist an interchange of electrons: these are known as insulators. Others that lie between the two are called semi-conductors.

Our interest lies in the ability of these moving electrons to do work, and a very simple example

is the illumination of a bulb used in a quartz watch. In the same way that rubbing your hands together generates heat through friction, so electrons flowing along a conductor generate friction and heat. The friction is so great through a fine conductor – the bulb element, that is – that it glows white-hot, thus providing illumination of the display.

ELECTRICAL UNITS

In a simple electric circuit there are three things to be measured: current, electromotive force, and resistance.

Electric Current

The unit used to measure electric current is the ampere, denoted by A or I, which is 1.6×10^{19} electrons passing a particular point in one second, which can be compared to litres of water coming out of a pipe in a given time. Water is often used as an analogy for electricity because the behaviour is so similar. (Fig. 140)

Electromotive Force

For current to flow between two points, a potential difference must exist between them. For example, a good battery is ionized so there is a surplus of electrons at one battery terminal and a deficiency of electrons at the other – so that, given a conductor, electrons will flow to balance the charge.

Just as water flows at a rate according to pressure, so too will the rate at which current flows depend on pressure. Electrical pressure and

Fig. 140 Litres of water coming out of the end of a pipe can be compared to the amp which is 1.6×10^{19} electrons passing a point in one second.

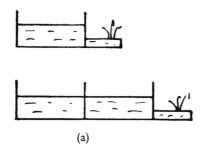

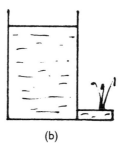

(a)   (b)

Fig. 141(a) More water doesn't necessarily give greater pressure. (b) A greater head of water gives greater pressure.

potential difference is measured in volts, the symbol for which is 'V'. One volt is the result of a current of one amp flowing in a circuit, giving one ohm resistance (*see below* for explanation of 'ohm'). (Fig. 141)

Resistance

As you might imagine, the pipe that water flows through offers a certain resistance – and the smaller the diameter of the pipe, the greater the resistance and the less water that can flow through it. Conversely, the larger the diameter of the pipe, the more water that will be able to flow through. The position is similar with electric current. The unit of electrical resistance 'R' is the ohm denoted by the symbol Ω. One ohm resistance is the result when a pressure of one volt is applied at one amp. This brings us to Ohm's law, which states:

> The current passing through a wire at constant temperature is proportional to the potential difference between its ends.

From this we can see that if any two of the units we have looked at are known, the third can be calculated.

$$V = I \times R \qquad I = \frac{V}{R} \qquad R = \frac{V}{I}$$

A simple way to recall the various formulae is to imagine a three-core cable with V, I and R written where wires might have been. (Fig. 142)

Fig. 142 An aid to remembering the basic electrical formulae.

Covering the V should remind you that $V = I \times R$, covering I should give you $I = \frac{V}{R}$

and covering R becomes $R = \frac{V}{I}$

Multiples and Sub-Multiples

Both in general and in horology, the quantities used to measure electrical units can be rather large or small, involving awkward numbers. For example, the unit of capacitance is the farad, symbol F, and the trimmer of a quartz watch may be 0.000000000028F. (The farad is the ratio of the charge to its potential, but in horological repair work there is virtually no need to know this.) To avoid such awkward numbers, whether large or small, prefixes are used to denote $\frac{1}{1,000}$ of something or 1,000 times something. The universal convention is:

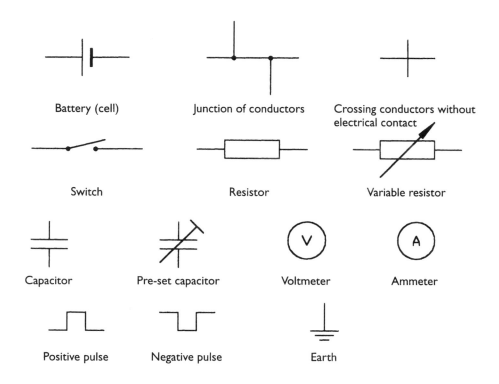

Battery (cell)   Junction of conductors   Crossing conductors without electrical contact

Switch   Resistor   Variable resistor

Capacitor   Pre-set capacitor   Voltmeter   Ammeter

Positive pulse   Negative pulse   Earth

Fig. 143 Electrical and electronic graphical symbols.

		Symbol
mega	$= 10^6$	M
kilo	$= 10^3$	k
actual unit (down to 0.999 or up to 999.0)		
milli	$= 10^{-3}$	m
micro	$= 10^{-6}$	μ
nano	$= 10^{-9}$	n
pica	$= 10^{-12}$	p

Using this convention, 0.000000000028F becomes 28pF: a much easier number to handle.

To give you some idea of the scale of units used in quartz watch repair, here is a comparison with a domestic power supply: as we know, electric current is measured in amps. The supply to a domestic cooker is usually rated at 30A and domestic lighting 5A. With a supply voltage of 240V, any of these has the potential to kill. For a watch, we are normally dealing with between 1.35V and 3V at a much lower amperage.

Watch consumption is measured in μA, and a consumption of 0.2μA to about 4μA is very common. It is also worthy of note that where the voltage and current in a quartz watch should present no dangers to the repairer, the static electricity in a person could damage the circuit of a quartz watch!

Symbols and Measurements

Electrical and Electronic Graphical Symbols
For electrical work, symbols are used to represent three-dimensional objects in two dimensions; for example when drawing a circuit diagram. Fig. 143 shows those commonly used in horology.

Meters for Measuring Electrical Quantities
Volts, amps and ohms are measured on a multi-meter. For general use in the workshop, an inexpensive multimeter could be used, but for quartz watch repair, a dedicated meter that can display microamps (μA) is highly desirable.

Voltage Measurement

A voltmeter is placed in parallel across the unit being measured; it is a high resistance instrument, meaning that it will take little power from the circuit being measured.

Ammeter

These are always placed in series with whatever is being measured, and are low resistance instruments to avoid introducing an extra, unwanted resistance.

When measuring resistance, if the coil cannot be isolated from the rest of the watch, make sure that the test voltage is no more than 0.2V otherwise damage can be caused to the watch.

11 QUARTZ WATCH REPAIR

Often I think of horology as a slowly evolving activity marked occasionally by a discovery or an invention which may be described as remarkable or even dramatic, and which has a significant impact on time-measuring devices. Two examples of significant advances in horology were the discovery of the properties of the pendulum, and the use of a spring as a driving force for a clock or watch and as a restoring couple for a balance.

Early watches with a balance had quite low train counts (the train count is the number of vibrations per hour of a balance for a watch to keep good time); in the mid-1950s most of the watches passing through my hands had balances that vibrated at about 18,000 vib/hour. For a number of years from about that time, faster and faster trains were developed, using counts of 19,800, 21,600 and 28,800, until train counts of 36,000 vib/hour were achieved. However, it was soon realized that 36,000 vib/hour, or 10 vib/sec, was about as fast as could be achieved mechanically, and so the search began for a faster oscillator that would have even better timekeeping properties. One development was the tuning fork which vibrated at 360 vib/sec; another – which has already outlived the tuning fork significantly and appears to have some way to go yet – is the quartz crystal, which for watch work vibrates at or about 36,768 vib/sec: such watches are capable of keeping time to within about 10 seconds a year.

We know that throughout our horological history, telling the time is all about an event which is repeated over and over in the same space of time and then counted up. In the case of a pendulum, balance, tuning fork or quartz crystal we can think of each as a type of oscillator because each is cyclic in action. However, to understand more clearly why a quartz crystal can be, and is, used in a watch, we need to understand something called the 'piezo-electric effect' – which is quite simply explained by looking at how an electric cigarette lighter works.

Inside the lighter is a piece of quartz crystal which is 'deformed' each time the lighter operates. When this happens by striking or by compression, the quartz crystal generates a high voltage but low amperage electric current which is caused to flow along a conductor. This has a small break in it, so that if the current is to continue to flow, it has to jump the gap – and in the process it causes a spark, which ignites gas turned on during the process of deforming the crystal.

In the watch we have the reverse of this. A small battery powers an electronic oscillator that continually feeds the quartz crystal. This causes the crystal to vibrate at its natural frequency, which in turn controls the electronic oscillator by electrical feedback. Think of this in a similar fashion to the way an escapement keeps a balance vibrating by delivering regular impulses of just the right strength and at just the right time. The balance is dependent on an impulse to keep it going, but the timing of the impulse is determined by the natural frequency of the balance. Each is dependent on the other, and just as the frequency of the balance determines the timekeeping of a mechanical watch, so the natural frequency of the quartz crystal determines the timekeeping of a quartz watch.

There are other reasons why we use a quartz crystal apart from the piezo-electric effect. One is the low energy loss of the quartz crystal, a second is because of its temperature character-

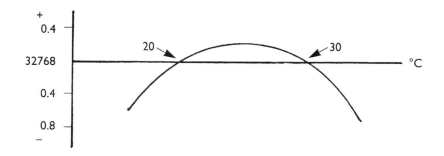

Fig. 144 Temperature characteristic of a non-compensated quartz crystal.

istic, and a third is its stability. Low energy loss refers to the self-damping properties of a vibrating quartz crystal, comparable to the desirability of a low frictional loss of the balance of a watch. For a non-temperature-compensated quartz crystal, the temperature characteristic is parabolic (Fig. 144).

It will be noticed that at wrist and room temperature – that is, the temperature most likely to be encountered by the watch – the crystal vibrates to frequency. When it is below room or above wrist temperature, the crystal frequency is a little slow, and when it is between the two, the crystal frequency is a little fast. In practice, the watch should keep time to closer than 3 minutes a year, and

as close as 10 seconds a year if temperature-compensated.

Quartz is silicon dioxide and is a naturally occurring substance, although in its natural state it has too many impurities to be used in a watch. Instead, the quartz crystal is grown artificially in a laboratory and cut for the job it has to do. Originally quartz crystals for watches were bar shape, with two contact wires secured to two immovable points on the crystal called nodal points. Currently crystals are shaped like a tuning fork and sealed in a metal container called a can with two leads, one connected to the integrated circuit (IC or chip), the other to the trimmer. The function of these will be explained later. (Fig. 145)

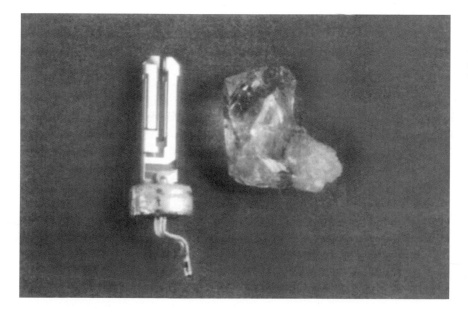

Fig. 145 A naturally occurring quartz, and a quartz crystal grown in a laboratory for use in a watch.

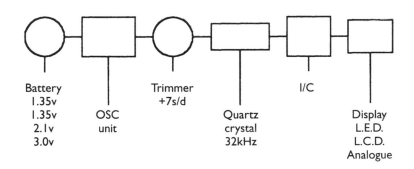

Fig. 146 Schematic diagram of the components of a quartz watch.

THE COMPONENTS OF A QUARTZ WATCH

Before we analyse the overhauling of a quartz watch, we will first consider its various components; these are shown in schematic form in Fig. 146.

Motive Force

Most usually a quartz watch is driven by a battery of 1.55 volts, occasionally of 2.1 volts or maybe 3 volts; a few early quartz watches had a 1.35 volt battery. These are still available, although not every battery manufacturer makes them. Alternatively a watch's motive force may be secondary cells charged by solar power, capacitors as used on the Seiko Kinetic, or accumulators as used on the ETA 205.911 Auto Quartz.

Oscillator Circuit

This is the second component, although it is not recognizable in the watch because it is part of the IC.

Trimmer

Thirdly we have the trimmer which is able to 'pull' the frequency of the quartz crystal to the extent of plus or minus seven seconds a day. This is possible because the electrical behaviour of the crystal is largely capacitive and the frequency can be adjusted by a variable capacitor. The latter is set by the manufacturer or

subsequently the repairer, and not by the customer.

The effect of adjusting the trimmer can be shown graphically. The curve showing the change in rate is sinusoidal rather than linear, as in a mechanical watch. Often there is no way of knowing which way to turn the trimmer – clockwise or anticlockwise – to effect a slight gain or loss, nor is it predictable by how much it has to be turned for a particular change of rate. The graph in Fig. 147 shows the change for a complete turn of a trimmer.

Not all watches have a trimmer, and in fact many early quartz watches had a fixed capacitor instead so the rate could not be altered. Later, alternative regulation systems were developed including predictable types, and a factory-set type of regulation system (see p. 139).

Quartz Crystal

As has already been said, the quartz crystal vibrates at 32,768 vib/sec (2^{15}); currently it takes the shape of a tuning fork and is vacuum sealed in a metal can. The manufacturer will have carried out a pre-ageing process by heat treatment; also he would have adjusted the frequency of the quartz. One technique is to tip the crystal with gold plating, then to vaporize the plating by laser until the frequency of the quartz crystal is within the desired control of the trimmer.

Provided the can isn't punctured, the quartz crystal should present no problems over the life

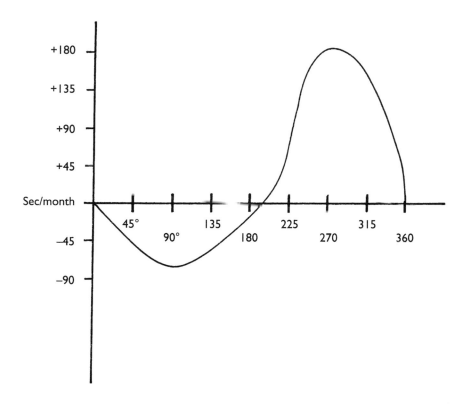

Fig. 147 The effect on the rate for one turn of the trimmer.

of the watch. Very occasionally a soldered connection to the quartz becomes loose, but resoldering restores the watch again.

The IC

The brain of the watch is the IC: it houses the oscillating circuit, and it is also where 32,768 is divided down to give an output to feed a display. It is one component which can never be repaired, and a replacement is never available: in short, it is part of the circuit board, and if a fault other than a bad solder connection is found, a whole new circuit board has to be fitted. There could be thousands of transistors in the IC which itself measures only a couple of millimetres square, so given its size and its complexity, a repair would not even be considered.

The IC is normally covered with a ceramic material for protection. In early quartz watches it has eight connections (sometimes called outputs): two connected to the quartz crystal, a third to the battery, a fourth to earth, two feed the display of an analogue watch, and, on a centre-seconds watch, one forms part of the grounding switch which grounds (not shorts) the output from the IC so that the seconds hand remains stationary during hand-setting. The eighth lead is not used.

The Display

The display may be analogue – that is, with hands just like a mechanical watch – or it may be a liquid crystal display (LCD) (*see* p. 142). The earliest quartz watches had a light-emitting diode (LED) type display, but this was phased out for the following reasons: (a) it had a high consumption, so high that a continuous display wasn't practicable; (b) it was difficult to read the

Fig. 148 Circuit diagram of a
quartz watch.

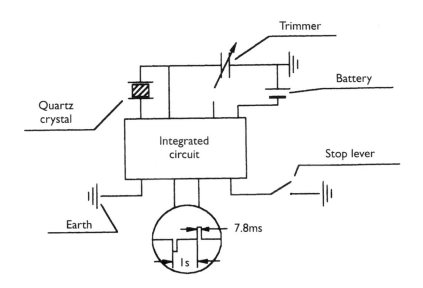

time in bright sunlight; (c) both hands were needed to operate a push button to display the time. None of these problems exist with analogue or LCD.

The output from the IC shown in Fig. 144 is described as 'alternate positive and negative impulses'. These are at one-second intervals of 7.8 milliseconds (ms) duration in our practice watch. What it means is that current flows at one-second intervals first in one direction for

7.8ms, then in the other direction for 7.8ms. (The duration of the pulse can be slightly different in different watches.)

The Stepping Motor

For most of the time the battery feeds the IC, trimmer and quartz crystal with a small operating current, and there is no drain in the current caused by the stepping motor except for the brief moment of impulse once a second for 7.8ms.

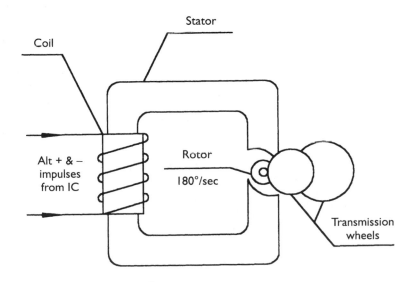

Fig. 149 A
stepping motor.

The coil is connected to the output of the IC, and is in very close proximity to the soft-iron stator. Between the open ends of the stator is a rotor which is a permanent magnet with two poles: a north pole and a south pole. While no current flows, the rotor aligns itself as shown in Fig. 150. When the IC sends out a pulse, the coil is polarized north/south. The soft-iron stator, which has lost any significant previous polarity, is energized by induction so that the north pole of the coil is transferred to one end of the stator, and the south pole of the coil to the other end.

So we now have the north pole of the stator adjacent to the north pole of the rotor, and the south pole of the stator adjacent to the south pole of the rotor. Because like poles repel, the rotor will turn, and it will turn in a particular direction due to the offset poles of the stator.

Soon after the rotor starts to turn, the magnetic field of the stator collapses due to the current being switched off by the IC, and the rotor runs on due to inertia and aligns itself

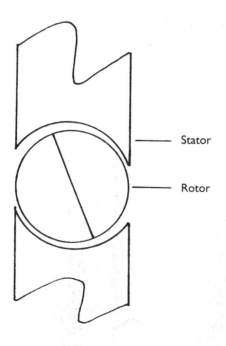

Stator

Rotor

Fig. 150 When no current is flowing, the rotor aligns itself symmetrically about the closest point of the stator.

again with the stator, which has now lost virtually all its magnetism. One second after the last pulse, again the IC sends out a further pulse which, as always, is the reverse polarity to the last pulse so that with each pulse the rotor turns 180°. The rotor has a pinion which drives transmission wheels to advance an hour and a minute hand, and very often a centre seconds hand, too.

OVERHAULING A QUARTZ WATCH

Before a quartz watch is overhauled, certain tests are generally carried out to establish whether faults are mechanical or electrical. However, before explaining these tests we will first strip and reassemble a quartz analogue watch, largely to familiarize you with all the working parts, some of which are still to be identified. The watch I have chosen to dismantle is an ESA 940.111 because it is very representative of many quartz analogue watches, and what is learned is therefore largely transferable to other watches. (Fig. 151)

It is highly desirable that you use antimagnetic tweezers for handling components of quartz watches. Ordinary steel tweezers will inevitably become magnetized and will attract screws and other small steel components, and this is very annoying. In addition the rotor will cling to them, making reassembly unnecessarily difficult. If you cannot get hold of antimagnetic tweezers, use brass ones, as rotors are not attracted to these, either. It is useful to keep a simple demagnetizer in the workshop, not only to demagnetize tweezers after handling quartz watches but mechanical watches, too. Never demagnetize a complete analogue quartz watch, and in particular never demagnetize the rotor of a quartz watch. To demagnetize small parts such as screws, simply wrap them in tissue paper, hold them close to the coil of the demagnetizer, switch on, and then withdraw the parts to arm's length.

Another invaluable, low-cost tool for quartz watches is plastic tweezers for handling batteries, as they help prevent accidentally shorting a

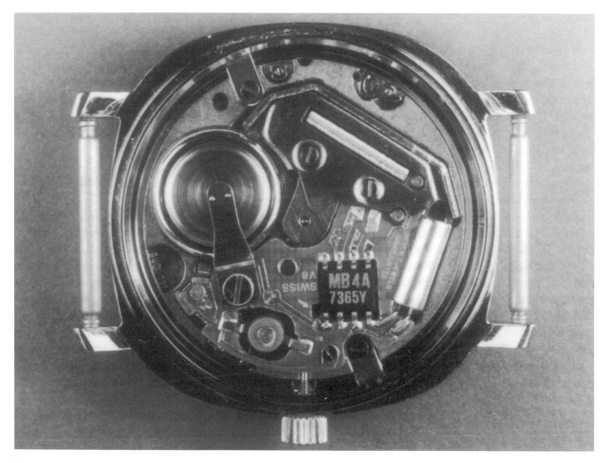

Fig. 151 A typical quartz analogue watch.

battery. It is possible to get antimagnetic screw-drivers, too, but personally I find them too soft and they increase the likelihood of slipping and causing damage.

Dismantling

1. Take out the movement from the case and remove the hands, dial, hour wheel, battery clamp and battery; then put the movement into a special holder and remove the upper magnetic screen: this protects the rotor from stray magnetic fields associated with domestic electrical appliances. (Fig. 152)

2. Remove the three screws holding the printed circuit board, one of which is coloured black or blue. Great care is needed when removing the coloured screw close to the coil as it is only too easy to slip with the screwdriver; also, be careful not to use too large a screwdriver as you could very easily damage the coils by touching them with it. This is a real danger even for the experienced worker.

3. Lift off the electronic module: underneath is a blue distance piece which may remain on the watch, or may come away with the module. It is quite safe to lift it off the module. Both of these will be cleaned with isopropyl in the form of injection swabs.

4. Remove the blue distance piece if it stayed with the module.

Fig. 152 Upper magnetic screen removed.

Fig. 153 Showing the stop-second lever and the underneath side of the electronic module with the post of the grounding switch.

With the train bridge removed, not only will you see the rotor and train but you will also see the stop-seconds lever. It has three legs: one engages with the sliding pinion, the centre leg is part of the grounding switch, and the third leg is the stop-seconds lever. With the stem in position 3 (fully out), it presses lightly against the second wheel (centre-seconds wheel) so that it remains stationary while hand-setting. (Fig. 153)

5. Remove the intermediate wheel (driven by the rotor), the second wheel, the third wheel and the rotor. All but the rotor can be placed in the cleaning machine. Keep the rotor away from other steel components.

6. Remove the stop-seconds lever.

7. Push the post of the setting lever and remove the stem and sliding pinion.

8. Turn the watch over and remove the calendar work and the motion work.

While the rest of the watch is in the cleaning machine, clean the rotor with Rodico. Rodico is particularly good at removing small steel particles picked up by the rotor. At the same time, clean the electronic module, the distance piece and the negative battery clamp with isopropyl.

Reassembly

1. Inspect the jewel holes in the bottom plate in the usual way.

2. Reassemble the hand-setting mechanism, lubricating with pressure-resistant oil or grease as you go, just to make sure nothing is missed. My personal preference is for watch grease. Replace the date-corrector operating lever, the setting lever, the yoke and the setting-lever jumper. Secure with its screw and load the setting-lever spring behind the setting lever and the yoke spring, which is part of the setting-lever jumper, behind the yoke.

3. Put the bottom plate in the movement holder and replace the sliding pinion, making sure it is engaged with the yoke and with the teeth pointing towards the centre of the watch.

4. Grease the stem and, from the outside of the plate, slide it into the movement and into the sliding pinion. The setting lever should move back of its own accord and relocate once the stem is in position. Test the hand-setting mechanism: it should have a positive feel about it for all three positions. Sometimes the setting-lever jumper becomes disassociated from the setting lever, in which

case, relocate it. Grease the groove in the sliding pinion. Make sure the stem is returned to the normal operating position.

5. Replace the stop-seconds lever, making sure that it is located in the groove in the sliding pinion.

6. Replace the third wheel, feeling the bottom pivot into its hole (pinion down).

7. Oil the shoulder of the second wheel (centre-seconds wheel) where it bears against the post in the bottom plate, and replace.

8. Replace the rotor with its pinion facing up.

9. Replace the intermediate wheel, pinion down. Again, you will have to feel the pivot into position.

10. Having inspected the jewel holes, replace the train bridge. I usually place the bridge into position, replace the train-bridge screws to stop the bridge from sliding around but without tightening them, then manipulate the pivots into their holes. Do this by placing your thumb under the movement holder while holding the plate down with tissue paper between the bridge and your finger. Now manipulate the wheels with a fine oiler or pricker: like this there is less chance of breaking a pivot. If necessary use a double eyeglass so that you can see and feel pivots into place.

11. When you are satisfied that all the pivots are in place, tighten both countersunk train-bridge screws.

12. The far side of the train bridge will eventually be held down by the same screw that holds the upper magnetic screen, and it is quite common at this stage of reassembly for the train bridge to be lifted slightly. Hold the bridge down with your fingernail while you test for endshake in the train. The rotor of course will not lift and drop under its own weight due to its magnetism, but you should still be able to lift it with your oiler.

13. Check the operation of the stop-seconds lever.

14. Oil all the pivots in the train bridge with the lightest grade of watch oil available – Moebius 9010 is just about light enough,

Moebius 9030 is likely to be even better as it is a light oil intended for cold temperatures.

15. Remove the movement from the holder and lubricate the train bottom pivots in a similar way to the top, including the front pivot of the second wheel.

16. Grease the friction clutch of the cannon pinion and replace it on its post.

17. Grease the posts of the minute wheel, setting wheel, intermediate setting wheel and the pivot of the date corrector.

18. Replace the minute wheel, setting wheel, intermediate setting wheel and the date corrector.

19. Replace the minute-train cover and its screw, making sure the eyelet is properly located under the plate. Check the operation of the parts just replaced.

20. Grease the bottom pivot of the date-indicator driving wheel, and replace.

21. Replace the date indicator, making sure the teeth go under the minute-train cover.

22. Grease the post of the intermediate date wheel, then pass the smaller wheel of the intermediate date wheel through the hole in the date jumper, and replace both together,

locating the jumper with respect to teeth in the date indicator.

23. Replace the jumper-maintaining plate and its screw.

24. Grease the jumper and a number of teeth around the date-indicator so that all teeth will be lubricated as the watch works.

25. Lubricate the outside of the cannon pinion with watch grease – but not where the minute hand fits – then replace the hour wheel and check the functions of the hand-setting and calendar mechanism.

26. Replace the centre-seconds hand in anticipation of electrical tests.

27. Turn the watch over, place it in the movement holder again and replace the distance piece.

28. Check that with the stem in positions 1 (pushed in) and 2 (the first pull out), the contact arm of the stop-seconds lever cannot be seen through the hole at the bottom of the distance piece, but that with the stem in position 3 (fully out), the contact arm *can* be seen coming across the hole. Remember to push the stem back to position 1 after this test. This is vital for the correct functioning of the grounding switch. (Fig. 154)

Fig. 154 When the stem is in position 3, the grounding switch should be seen coming across the hole in the distance piece.

29. Replace the electronic module, putting the coloured screw at the end of the coil.
30. Replace the upper magnetic screen.
31. Carry out appropriate electrical tests, as described below.

ELECTRICAL TESTS

The following electrical tests were carried out on an ESA 940.111, and are typical of those normally carried out on similar watches. The figures in brackets are the actual figures obtained on a demonstration watch. If, on another watch, the coil can be isolated from the IC, the probe position for resistance measurements is at the two ends of the coil. The tests that are usually done on quartz analogue watches are: consumption, operation of the grounding switch, lower working voltage limit, resistance of the coil, consumption of the coil, coil insulation, confirmation of positive and negative impulses from the IC, and finally rating. The expected values are quoted in manufacturers' technical information, and these should be obtained from manufacturers where possible. They also appear in the Flume Electronic Service System available from material houses.

To do all of the above tests does call for expensive, dedicated electrical test equipment, and could be beyond the budget of even professional repairers – its acquisition certainly couldn't be justified by most enthusiasts. Fortunately, many of the tests can be done on lower cost equipment, and since the rate of many watches is fixed by the manufacturer, there isn't the same need to check and adjust this.

The test equipment used here to complete most of the common electrical tests is the Witschi Q Test 6000; however, I shall suggest low-cost alternatives where they are available and known to me.

Consumption
A quartz analogue watch with a seconds hand

Test	Remarks	Measurement	Probe position
Consumption of the movement	Measurement without the battery, with an external current supply of 1.55V. Stem pushed in	$<6\mu A$ $(3.4\mu A)$	At the positive and negative battery contacts
Grounding switch	Measurement without the battery, with an external current supply of 1.55V. Stem in pos. 3	0.2 to $0.3\mu A$ $(0.2\mu A)$	At the positive and negative battery contacts
Lower working voltage limit	Measurement without the battery, with an external current supply of 1.55V. Stem pushed in	The watch should continue to function as the voltage is dropped down to at least 1.3V	At the positive and negative battery contacts
Coil resistance	Measurement without the battery	$2k\Omega$ $(2.19k\Omega)$	At the output of the IC
Alternate positive and negative impulses from the IC	Test on Witschi or Etic Microtest 2000	LED's flash alternately	Probes not necessary. Pick up on sensor

should have a comparatively low consumption for most of the time, peaking for the duration of the impulse. This is because for most of the time only the IC, the quartz crystal and the trimmer are consuming power. However, for a period of about 7.8ms during each second the coil is fed, and this consumes significantly more power in addition to that consumed by the IC, crystal and trimmer. It is therefore necessary to establish the watch's average consumption over a complete second (or over 5 seconds for a watch without a seconds hand that impulses every 5 seconds, or for any other given period – 10 and 20 seconds being quite common).

Most of the tests are carried out with the seconds hand fitted. The watch is held in a movement holder over a mirror, and by looking in the mirror it can be seen whether the watch is working or not by observing the hand. This of course can't be done if the watch doesn't have a seconds hand, in which case try to see the rotor or one of the faster-moving train wheels to check that the watch is working.

The test is carried out with the battery removed, and with the test equipment itself supplying a current at the normal working voltage of the watch; this supply is referred to as an 'external current supply' and is fed to the normal battery connection. The stem must be pushed in for this test so that it is in the same position as when the watch is worn.

Like many other watches, the 940.111 has calendar work as well as the facility to tell the time, and it has a three-position stem. Fully pushed in, at position 1, the button and stem have no function. At the first pull out of the button, position 2, the watch continues to function normally but offers a rapid change of the calendar. The second pull out, position 3, grounds the output from the IC for hand-setting. In position 3, the calendar advances once every twenty-four hours at about midnight, the same as a mechanical watch.

A significantly less expensive meter for measuring the average consumption is the Seiko S-840A Digital Multi-Tester with MA-40A Multi-Adapter. A problem in using an ordinary micro-ammeter is that it cannot average the consumption between pulses.

Operation of the Grounding Switch

A watch with a centre-seconds hand is likely to have a grounding switch so that the hands can be set and synchronized together and with a master clock or time signal. During hand-setting, the IC, quartz and trimmer continue to be fed but there is no output from the IC to feed the coil, therefore the consumption will drop significantly. A drop to 10 per cent of the original value or less is quite common so that a watch with an average consumption of, say, 2.8μA could well drop to 0.2μA when the grounding switch is operated by pulling the button out to position 3. The feed is as for the consumption. The measurement could also be done on the Seiko equipment.

Lower Working Voltage Limit

We need to establish that the watch will continue to function even when the battery voltage begins to drop. The test also helps to establish that there is no significant frictional loss in the mechanical parts, caused for example by dry oil.

The test is carried out with a variable external current supply feeding the watch at the battery supply, with the stem pushed in to position 1. A centre-seconds hand needs to be fitted for this test particularly so that as the voltage is lowered, the actual point at which the watch stops can be observed. I would expect a watch that normally operates at 1.55V, to continue to function at 1.35V. Features of the watch not yet mentioned may come into play at or below this figure, so don't be too surprised if the watch functions differently with a voltage drop (for example, rapid pulses to the seconds hand, followed by a brief period of inaction).

Coil Condition: Resistance, Consumption and Insulation

Electrical resistance is measured in ohms (Ω), and in the instance of watch coils, is usually in thousands of ohms, written kΩ and pronounced kilohms. Typical values are between about 1kΩ and 3kΩ.

Ideally the coil should be isolated from the IC so that the voltage being fed to the coil doesn't back-feed the IC and cause damage. Test equipment dedicated to watch work should be safe to use, even if the coil can't be isolated from the IC, but if a more general test meter is being used, the test voltage could just damage the coil or the IC. Ohmmeters with a test voltage higher than 0.4V are unsuitable for watch work; recommended is a test meter of 0.2V.

Set your test meter up to read resistance so that you can read about 2kΩ. Feed the ends of the coil without regard for polarity, and accept a reading within about 10 per cent of the nominal stated resistance.

The Seiko equipment can be used for this and the coil insulation test.

Coil Consumption
When the coil cannot be isolated from the IC, instead of testing for its resistance, its condition can be gleaned by subtracting the consumption in μA, obtained with the stem pulled out, from the average consumption with the stem pushed in.

Coil Insulation
Simply set the equipment up for measuring resistance, then connect one end of the meter to one end of the coil and the other lead to earth (the metal part of the movement). You should find an open circuit: that is, an infinitely high resistance.

Alternate Positive and Negative Impulses
A very useful and inexpensive instrument for this test is the Etic Microtest 2000. The movement doesn't even need to be removed from the case. Simply switch on the tester and place the cased or uncased watch on the sensor. An audible blip accompanied by alternately flashing green diodes indicates that there is an output from the IC and that the coil is in order. An audible blip but no flashing diodes, or just one, suggests the coil is not functioning but that there is an output from the IC.

RATING

If there is a need to rate the watch, unfortunately it does require the use of an expensive instrument. There are four signals from the watch that could be used to check the rate: they are capacitive, magnetic or inductive, acoustic and current drain.

Capacitive Rating Signal
This signal is from electrical stray fields from the quartz oscillator or, in the instance of LCD, the operating frequency of the digital display.

Magnetic or Inductive Signal
The sensor detects stray magnetic fields from the coil of the stepping motor.

Acoustic Signal
In a mechanical watch, the unlocking signal is used which includes the impulse pin striking one side of the notch and the unlocking of the escape wheel. In a quartz watch, the mechanical vibrations of the quartz are detected.

Rating by Current Drain
Here the signal is derived from the supply current, or by inserting a sensor between the battery and its contact.

Before rating a quartz watch, remember that some crystals are temperature sensitive, and if a customer takes a watch off for regulation, the instantaneous rate will be changing as the watch passes from wrist to room temperature. It would be better to wait for half an hour for the watch rate to settle at room temperature, then rate the watch. During this time, have the rate recorder switched on so that the internal crystal has time to stabilize.

Having selected the right signal to detect on the rate recorder, place the watch with an open back on the pick-up, if necessary moving the watch around so as to find the most favourable position to pick up a good signal. A rate of 0.16 seconds a day corresponds to about one minute a year, which is the same as 4.8 seconds a month.

Tools for adjusting the rate are inexpensive, but not really necessary. I often use part of a plastic knitting needle with an end shaped to fit the trimmer which is either a screwdriver slot or a square hole. Avoid pressing on the trimmer as adjustments are made, as they are often unsupported and pressure could cause damage. Often there is no way of knowing which way to turn a trimmer or by how much: in such instances just use trial and error.

DEVELOPMENTS
END-OF-LIFE INDICATION (EOL)

Many watches with seconds hands are able to give advance warning of impending battery failure. Every 60 seconds there is feedback from the battery to the IC, and if the battery is nearing the end of its life, the IC will put out four rapid pulses followed by no pulses for 4 seconds, then repeat the pattern again. This comes into operation when the battery voltage drops to 1.4 volts, giving the user of the watch about one week's notice of battery failure. (Fig. 155)

When replacing a faulty battery, you need to wait for up to a minute for the watch to resume normal working.

Rapid Pulses

Not all watches have a seconds hand, or impulse each second. Common alternatives are watches that impulse every 5, 10 or 20 seconds. To establish the average consumption of these watches, the test equipment needs to be set for the time between successive impulses, a period referred to as the measuring time, or sometimes the integration time or the gate time. If the measuring time is not pre-set but left by default at one second, you would read a very low consumption for most of the time, then the consumption would peak at the moment of the impulse.

With such watches, it is difficult to establish quickly that there is an output from the IC. Usually one looks for the rotor to see if it turns, but even then, an impulse can be missed. ETA overcame the problem with a facility for 'rapid pulses'. On the calibre 255.411, bridging a test point 'T' with a marked negative terminal, sixteen pulses a second are delivered and so the train wheels advance at an accelerated rate to show that the watch is functioning. On this calibre, if the voltage drops to 1.4V, the threshold for EOL, there will be thirty-two pulses per second. The rapid pulses are of more significance on the ETA 255.441 which impulses every 12 seconds. To activate rapid pulses of thirty-two per second, connect 'T' to the marked positive point. (Many ETA watches have a combined 'switch' and 'test point'. All such calibres will give rapid pulses if bridged to negative. Some calibres will need 'T' bridged to the frame of the watch for rapid pulses.)

When working normally with a battery voltage of 1.55, the consumption of the 255.411 (also the .111, .121 and .122) should be equal to, or

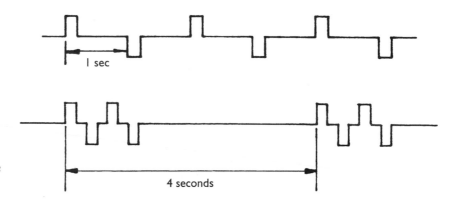

Fig. 155 Shows normal IC output, and the output when the voltage drops to below 1.4V (EOL).

less than 1.5µA. When EOL comes into play, expect a higher consumption. For the calibre 255.441, the average consumption should be equal to, or less than 0.7µA. All of these calibres should have a consumption of less than 0.5µA with the stem pulled out to position 3. Coil resistance for these watches is between 3.5 and 4kΩ.

Naturally, the consumption of a watch as it pulses thirty-two times a second will be exceptionally high. If it is supposed to pulse every 20 seconds the test meter will show 640 (32 × 20) times the normal consumption.

Although not necessary, other ETA calibres with seconds hands may still have the ability to give rapid pulses.

Power Reduction

Stepping motors in early ETA watches received a 7.8ms pulse which gave sufficient torque (torque is a force applied tangentially to a circle, but here it simply means power) to advance not only the rotor and train, but also to drive the calendar work. In fact this torque isn't necessary except perhaps during the 1½ hours that the calendar work is being driven. As the torque is dependent upon the duration of the pulse, it offers the opportunity of a shorter duration for each pulse for most of the day, and simply to increase the duration for driving the calendar if it is needed. ETA calibres 927, 255, 256, 955 and 956 are equipped with a new IC to accommodate impulses of different duration.

In practice, in the calibre ETA 955.412, normal pulses are 4.8ms each second, but if and when an extra load is put on by the calendar, if it is necessary, the duration of the pulse is increased to 7.8ms during the same second. The IC can detect if the rotor didn't turn. Once the IC is happy that the rotor is turning, the duration of the pulse drops back to 5.8ms for one minute, then reverts back to normal 4.8ms pulses. Overall a 30 per cent power reduction is achieved. Fig. 156 shows how a watch that stopped momentarily could be represented. The power saving is shown by comparing the consumption of an early watch without 30 per cent power saving, with that of a watch made after about 1983/4 which does have 30 per cent power reduction. Both watches have a 1.5V supply to a watch with a total resistance of 5kΩ.

Peak consumption $I = \dfrac{1.5V}{5k\Omega} = 300\mu A$

Average consumption of an older watch in
$\mu A = \dfrac{300\mu A \times 7.6ms}{1ms} = 2.34\mu A$

Similar watch with 30 per cent power saving
$I = \dfrac{300\mu A \times 4.8ms}{1ms} = 1.44\mu A$

Electrical Tests for the 955.412

- At 1.55V, the average consumption should be equal to, or less than 1.5µA with 1-second pulses.
- At 1.55V, bridging 'T' to negative gives you sixteen rapid pulses a second.
- At equal to or less than 1.4V, bridging test to negative gives you thirty-two rapid pulses a second.

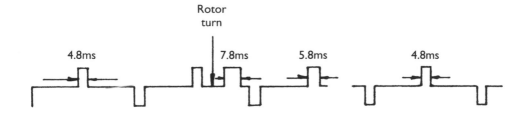

Fig. 156 Represents a stopped watch.

Fig. 157 A watch with an inhibition system of regulation.

- At 1.4V there should be four impulses every fourth second.
- At 1.55V the consumption with a mechanical fault could be between 4.4 and 13.2μA.

Some watches, including the ESA 955.412, have a 'reset' point 'R' printed on the circuit board, to be bridged with a positive contact. This is to return the watch to normal after EOL to save waiting for one minute for normal function to return naturally. The same can be achieved by pulling the stem out and pushing it in again.

INHIBITION

(Digital Frequency Regulation)

Prior to 1973, regulation of quartz watches was by capacitor, which pulled the frequency of the crystal up or down for the watch to keep good time. After 1973, ETA introduced, for their watches, a system of regulation based on a quartz crystal that always vibrates at one, two or three vibrations a second faster than 32,768 vibrations per second. By alteration to the dividing stages within the IC, the motor impulses are adjusted to the order of $\frac{1}{1000}$ of a second every 10 or 20 seconds, which is not apparent at the hands.

There are five tracks on the circuit board offering thirty-one possibilities of correcting up to about 8 seconds a day by cutting the appropriate tracks. (Fig. 157.) The following paragraph shows how inhibition works; it is not important if you do not follow it.

One vibration error at 32,768 vibrations per second = 30.5μ seconds. After the first division stage (2 stage) we have 16,384 vibrations a second, and one vibration error equals 61μ seconds or 5.27 seconds a day. One vibration correction every 20 seconds accumulates to 0.000061 times the number of 20 seconds in a day. $0.000061 \times 4320 = 0.26$ s/d correction.

The printed circuit board has tracks which are circuits that may be cut to correct errors.

Track	Correction s/d	Possibilities
1	0.26	1
2	0.53	2
3	1.06	4
4	2.11	8
5	4.22	16
		31 Total

Fig. 158 A predictable regulation system with sprung arms.

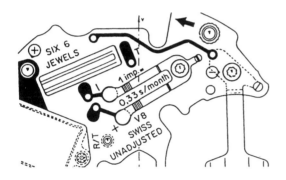

In practice, the manufacturer finds the error per day, then chooses which track or combination of tracks to cut to correct the error. For example, to correct a 5.28 second gain, tracks 3 and 5 would be cut. Cutting of the tracks is all carried out at the factory, and it is claimed that the effect of cutting tracks cannot be reversed. I haven't tried ... yet.

When testing the rate, the test must be inductive or magnetic, and not acoustic; that is, from the motor impulses. In addition to this, you must pick up the correct impulse – that is, impulses at 20-second intervals. Just set the measuring time for 20 seconds.

Yet another regulation system is used on ETA calibres 255.511 and 255.561, which is temperature-compensated to 10 seconds a year, has inhibition with a time period of 8 minutes, and EOL displays by five rapid pulses every fifth second.

Because the watch may be more accurate than the test equipment, ETA recommend timing the watch over a period of one month, then regulating the watch in 0.33 second steps by putting the stem into position 3 and pressing one of two spring arms on the back of the watch. (Fig. 158)

Fig. 159 Regulation by changing the capacitor.

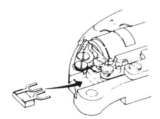

Fig. 160 Regulation by rotary step switch.

Cal. 23 Series:

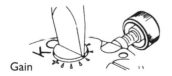

Gain

approx. 0.5 sec./day/step

Cal. 59 Series:

Rotary step switch for fine adjusting approx. 0.26 sec./day/step

Rotary step switch for rough adjusting approx. 1.04 sec./day/step

Calculate the number of presses by dividing the error in seconds over one month by 0.33. If the watch is losing, press the + arm; if the watch is gaining, press the −arm. It is only desirable to regulate if the error per month is greater than 0.8 seconds.

Seiko have two alternatives for regulating quartz analogue watches: one is to change the frequency of the oscillator circuit by changing the capacity of the condenser; the other is by use of a rotary step switch which is used to change the ultimate frequency of the frequency divider circuit. (Figs. 159 and 160)

Lastly, on some watches (and calculators) after fitting a new battery, it is necessary to bridge an 'all clear' (AC) point to the positive battery terminal or some other bridging point. This is essential for the watch to function properly.

12 LIQUID CRYSTAL DISPLAY

Before about 1979, when I first began servicing liquid crystal display watches, it seemed quite incredible that one could have a watch with no moving parts. Today most people have at least seen watches with liquid crystal display (LCD), and many rely on liquid crystal display somewhere in the home, be it for a watch, clock, time-switch, camera, thermometer, security alarm or tyre-pressure gauge.

Most LCD watches are associated with the lower end of the market in terms of cost, but they do have the advantage of additional functions such as chronograph, countdown, lap time, alarm and calendar, all at little extra cost and with easy-to-read display; yet they still have the same quality of timekeeping as quartz analogue watches. LCD watches do not have buttons, but pushers to select different functions and for setting the display. Inside the watch, the module – it is no longer called a movement – is much the same as a quartz analogue, with the exception of the display; in fact they are so similar that some watches have both analogue and LCD display, and share a common IC and quartz crystal.

LCD EXPLAINED

Earlier I spoke of the three states of matter: solid, liquid and gas (*see* p. 119). In 1888 a fourth state of matter was discovered, liquid crystal, which lay between a liquid and solid. In the mid 1960s, molecules of liquid crystal were modified by chemists so that they became fluid at room temperature and could be used commercially for display purposes. Of particular interest was the ability of liquid crystals to bend light, and the behavioural change to liquid crystal molecules when a voltage was applied to them.

Light can be thought of as a beam, such as comes from a torch, containing waves that

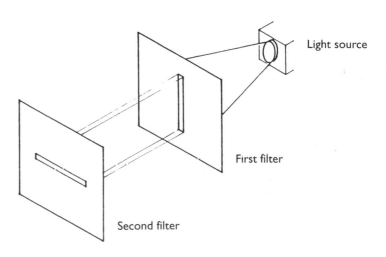

Fig. 161 Light passing the first filter is blocked by the second.

oscillate in many different planes. If filters are put in the path of the beam, some light passes the first filter, but it cannot pass the second. (See Fig. 161)

As children, perhaps most of us played with polarized sunglasses and observed that by placing one lens over the other in a particular way, they darkened significantly. This is because each lens is polarized in the same plane and allows only light from a particular direction to pass. When two polarized lenses are placed over each other so that the polar axis of one lies above the polar axis of the other, light passes. Rotate one of the lenses by 90°, and therefore one of the axes, and the light that passes the upper lens cannot pass the lower lens and so the overlapping lenses darken. (Fig. 162)

In a liquid crystal display watch, liquid crystal is sandwiched between two pieces of glass. Polarized film is attached to the outside of the upper and lower glass, but with their axes 90° opposed. For simplicity, let us assume that light

oscillates in two directions, north/south and east/west. Normally light enters the top polarized film from a north/south direction, but not east/west, so no east/west light even reaches the bottom polarized film. Because the bottom film is polarized in the opposite direction, if the liquid crystal were not there, the light that enters the upper film could not pass the lower film and so the display would appear dark.

However, the liquid crystal between the glass bends the light through 90° so it can now pass the lower polarized film and the display appears clear. This is possible because the two pieces of glass are separated by about twelve microns with rod-shape molecules filling the gap, and are aligned so that all the molecules of liquid crystal adjacent to the upper glass are parallel and aligned, say, north/south, while the molecules nearest the bottom glass are parallel and aligned east/west. The molecules between the two twist progressively from north/south to east/west. (Fig. 163)

Fig. 162 Two lenses with the polar axes parallel and the polar axes opposed by 90°.

Polar axis opposed by 90°

Polar axis parallel

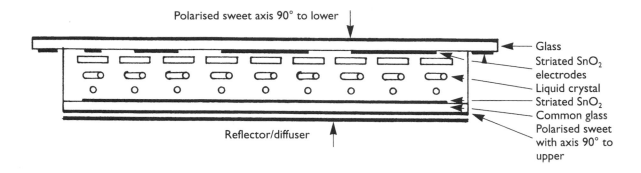

Polarised sweet axis 90° to lower

Glass
Striated SnO_2 electrodes
Liquid crystal
Striated SnO_2
Common glass
Polarised sweet with axis 90° to upper

Reflector/diffuser

Fig. 163 Molecules of liquid crystal twist through 90° between the upper and lower glass.

If we were now to coat the inside of the both pieces of glass with tin oxide (SnO_2) and apply a potential difference between the two pieces of glass, the molecules would up-end and would no longer twist the light – no light would therefore pass the lower glass, and the display would darken.

In a watch display, the bottom glass is coated with striated tin oxide – it has lines like a corrugated roof – to form a common electrode, and the upper glass is coated in segments so that when a potential difference is applied between certain segments and the common, only those segments darken and so we can display information. A display working on this principle is known as 'nematic rotation' or 'field effect', and it has an advantage over an alternative system called 'dynamic scattering' because the display consumes virtually no power, has a better working and shelf life, and has a good contrast for easy reading of the display. Each segment is connected to its own electrode on the inside of the upper glass, and when the display is removed from the watch, it is possible to see the segments and electrodes by catching the light in the right way on the display. (Fig. 164)

Fig. 164 By catching the light, the segments and electrodes may be seen.

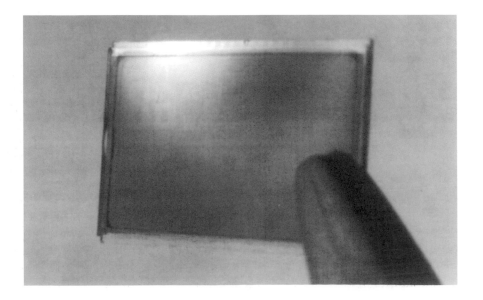

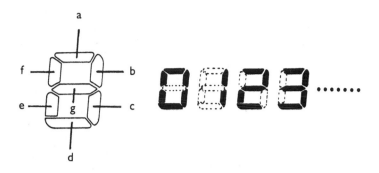

Fig. 165 Segments from one digit. For the figure eight, segments 'a' to 'g' are displayed.

In the simplest of displays there could be twenty-six electrodes: two for the figure one, seven for each of three full digits, two for a flashing colon and a common. This type of display, which has an electrode for each segment, is known as 'direct drive', and you may expect there to be thirteen electrodes on each side of the display. The segments from one digit are shown in Fig. 165.

A disadvantage with direct drive is the space that it takes up, and to overcome this, often a number of segments are given a common electrode, freeing space for other functions and at the same time making the watch more reliable – with fewer electrodes there is less to go wrong. This type of display is called 'multiplex drive'.

To provide a potential difference between the common and the individual segments, there are the necessary number of outputs from the IC – twenty-six in the example just given – which connect with the display via 'connecting strips'. Each of these – and there are usually two, but there may be only one – is a flexible material made up of a number of dark and light conducting and insulating wafers. When I look at connecting strips, I think of a bank of minia-ture liquorice allsorts joined together.

To make the display clearer, expect to find a diffuser in the space between the display and the module, which may be a separate membrane, or may be stuck to the printed circuit board. Because liquid crystal is not self-illuminating, there may be a lamp to assist with reading the display in the dark.

SERVICING AN LCD WATCH

There is very little to go wrong with an LCD watch, and very little to do other than fit new batteries. Sometimes it is possible to experience poor electrical contact for mechanical reasons, such as faulty pushers, or occasionally water gets into the watch which can cause electric current to pass between electrical paths that should be isolated from one another.

I use isopropyl for all the electronic module cleaning; I also carry out a consumption test of the module, often about 1μA, and of the lamp which is usually about 10mA, and I test the display and the alarm buzzer.

Remove case backs over the bench to prevent any loose parts being lost. This may not be necessary for all calibres, but some watches have contact springs that are not held captive that can be lost. Remember the orientation of the case back if it is shaped or a snap-on, as correct orientation may be necessary for the module to make contact with the alarm buzzer.

Overhauling the Seiko A914A
1. Remove the case back. You will see the battery clamp, the battery, and the buzzer-lead terminal which is a helical gold-plated contact spring: this one comes free if held with the tweezers and given an unscrewing action. (Fig. 166.)
2. Remove the battery clamp and battery.
3. Remove the module from the case.

Fig. 166 Inside the back of a Seiko A914A.

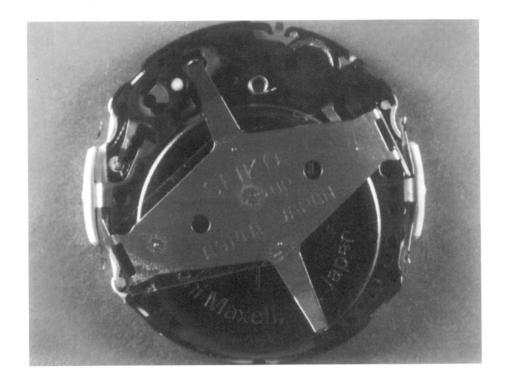

Fig. 167 The circuit board, trimmer, crystal and bulb can be seen.

4. Remove the circuit cover which carries the pusher contacts by unclipping it. The circuit board, trimmer, crystal and bulb for illuminating the display can be seen. (Fig. 167)

5. Carefully lift the green circuit block from the white framework. There are two location pins. As you lift the circuit block away, the display and connection strips will probably stick to it, and the reflecting mirror (diffuser/reflector) may slide out from under the display. (Fig. 168)

6. Carefully lift the display from the circuit block with plastic tweezers. The connecting strips will probably stick to the display: they usually do. Just peel them away. Parts of the Seiko A914A may be seen. (Fig. 169)

7. Clean the following parts with isopropyl:

Circuit cover	LCD panel frame
Battery clamp	Buzzer lead terminal
Circuit block	

Testing the Alarm Buzzer

Test the buzzer according to your test instrument. If an AC voltage of between 1.5 and 3V at 2kHz (hertz replaces cycles per second) is applied by placing one contact on the case back and the other on the piezo-electric element, the alarm will sound with the case back acting as a diaphragm. The piezo-electric element is glued to the case back. The amplitude of the sound is influenced by the voltage and is likely to sound quieter under test than it would do normally. In other watches the buzzer may be magnetic, in which case the alarm will sound as normal.

Testing the Display

If the display is direct drive, usually it can be tested with a 3V AC supply. Viewed from the inside of the display by means of a mirror image, the common is usually the bottom right-hand electrode. Position the negative flying test lead of your test instrument to the display's common, touch each electrode in turn with the red flying

Fig. 168 The circuit block, display and diffuser/ reflector.

Fig. 169
Parts of the
Seiko A914A

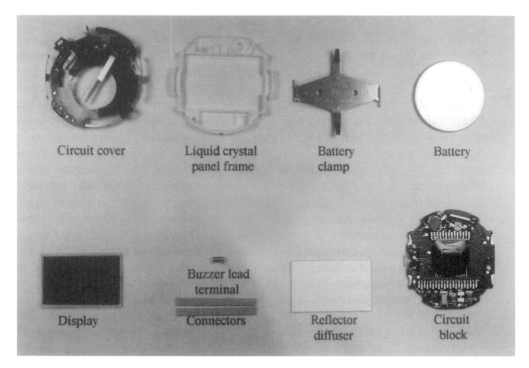

Circuit cover   Liquid crystal   Battery   Battery
                 panel frame   clamp

Display   Buzzer lead   Reflector   Circuit
           terminal   diffuser   block
           Connectors

test lead and the good segments should display. If a segment fails to display, a new display will be needed. It is a reflection of how reliable displays are that I have never had a display fail the test.

If the display is multiplex drive there will be more than one common, and electrodes will serve more than one segment. In the absence of a technical guide, I would suggest that by trial and error you observe all segments display. Alternatively, reassemble the watch and see if all segments then display.

To clean the display, I usually wipe over the polarized film with a clean soft cloth and gently rub Rodico across the electrodes. I do the same with the connecting strips.

When cleaning the case, remove the circlips from the pushers and clean inside them and the hole in the case. If necessary, replace the pusher seals and lubricate them with silicone grease before reassembly. Select a new gasket for the case back.

Reassembly
1. Replace the display into the liquid crystal display panel. Handle it with plastic tweezers as metal ones will readily chip it.
2. Having checked that there are no fine hairs on the display or the connectors, replace the two connectors. Invariably they stand upright.
3. Replace the diffuser/reflector.
4. Replace the circuit block.
5. Replace the circuit cover.
6. Replace the battery clamp.
7. Replace the buzzer terminal by 'screwing' the coil spring into its hole.
8. Carry out the electrical tests on the module. The average consumption should be about $1.7\mu A$, the consumption with the bulb illuminated should be about 8 to 9mA and the watch should be rated. Seiko claim accuracy to 20 seconds a month.
9. Having cleaned the case, replace the module.
10. Grease the gasket for the case back with silicone grease, and replace.
11. Dry-test the watch to three bars, set the hands and test all functions.

13 SERVICING A FIVE MOTOR CHRONOGRAPH

I have included this particular chronograph in this book for a variety of reasons. Certainly I feel that although many mechanical and quartz watches can be serviced without additional technical information by simply transferring knowledge from one watch to another, I would not have liked to have tackled this calibre and set the hands up without technical information.

The ETA 251.262 is a five motor quartz analogue chronograph which, in addition to showing hour, minute, continuous seconds and date, records ¹⁄₁₀ seconds, seconds, minutes and up to twelve hours. The hour hand can be advanced in one-hour jumps so it can be advanced or retarded without interfering with the rest of the watch. The button screws onto its pendant, and once unscrewed, has to be pulled out to position 2 to advance or retard the hour hand in one-hour steps. Position 2 also allows a rapid forwards- or backwards-moving date. Position 3 operates a 'stop-second switch' and allows hand-setting in the usual way. The date advances or steps back one date every twenty-four hours.

There are two pushers to operate the chronograph: one at the 2 o'clock position, which for convenience we will call pusher 'A', that starts and stops the chronograph; and the other at the 4 o'clock position, which we will call pusher 'B', that returns the chronograph to zero. There are other functions for the pushers in conjunction with the stem positions which correctly zero the hands and will be explained on completion of the watch overhaul. (Fig. 170)

Dismantling the ETA 251.262

1. Remove the case back. The five coils can be seen through the magnetic screen, the three lower coils coloured green: the one on the left is the 60-minute counter, the centre coil at the 6 o'clock position is for the time and looks after hours, minutes and continuous small seconds, and the coil on the right is for the 60-second counter. The top left red coil is for the ¹⁄₁₀ second counter, and the red coil on the right is for the 12-hour counter. (Fig. 171)

2. With the hands returned to zero, unscrew the button, and remove it and the stem by pushing on the end of the setting lever seen through the upper magnetic screen.

3. Unscrew the two case screws, turn the watch over, and the screws, clamps and movement will fall out.

Fig. 170 Five motor quartz chronograph, dial side.

9. Remove the six identical screws holding the black additional circuit board, and the board itself. Take great care with the screws close to the coils, because even experienced workers sometimes slip with the screwdriver and damage a coil. (Fig. 172)

10. Remove the red $\frac{1}{10}$-second coil and its screw (at top left, with the button on your left).

11. Remove the $\frac{1}{10}$-second train-bridge screw, the bridge, train and the stator. To make reassembly easier, draw the train layout. Use tweezers that the rotor isn't attracted to for handling the rotor. Though a little soft, brass tweezers are fine if kept dressed.

12. Remove the red 12-hour coil, the train bridge, the train, rotor and stator. Again, draw the train layout. You are now left with the three green coils.

13. Remove the near-elliptical, upper train bridge held with two screws, and the 60-second chronograph wheel which has five arms.

14. Carefully remove the lower train-wheel bridge with its two screws. Try not to disturb the train so that the layout can be drawn. (Fig. 173)

15. Remove the black and white circuit distance piece, noting which way up it goes.

16. Remove the three remaining coils and stators.

17. Remove the three tall-headed screws from the bottom plate, and the plate itself, exposing the electronic module. (Fig. 174)

18. Remove the 'cannon pinion with driver', the positive battery connection, and the one-piece stop-lever and switch.

19. Remove the stem and sliding pinion.

20. Remove the electronic module, taking the two contact springs for push-piece with it.

21. Make a drawing of the hand-setting and calendar mechanism, and remove the parts. (Fig. 175)

22. Check the resistance of the three green coils and the two red coils: that of the latter should be 1.5 to 2.5kΩ, and of the green coils 1.0 to 2.0kΩ.

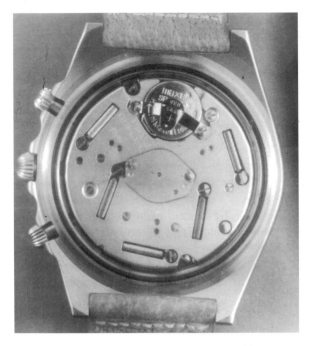

Fig. 171 Five motor quartz chronograph with the back removed.

4. Replace the button and turn the hands to a convenient position to remove them: this would be the continuous small seconds hand, the 12-hour-recording, the $\frac{1}{10}$ seconds-recording, the minutes-recording and the large seconds-recording hands, and the minute and hour hands. Cover with tissue paper in the usual way. The three small hands could be removed with the dial if they are too close to the dial to insert hand-lifters safely.

5. Remove the dial by unlocking the dial fasteners and lifting the dial.

6. On the dial side of the watch, remove the three identical, small date-indicator maintaining plates, and the date indicator.

7. With the movement on a holder, slacken the battery screws, remove the battery clamp and battery, and lastly, the two screws.

8. Remove the two screws securing the upper magnetic screen and the screen itself.

Fig. 172 The upper magnetic screen and additional circuit board removed.

Fig. 173 The main train with the bridge removed.

Fig. 174 The electronic module exposed.

Fig. 175 Hand-setting and calendar assembly.

Reassembly

This is very much the reverse of dismantling, except that you need to lubricate as you go.

1. Start by reassembling the hand-setting and calendar mechanism. Grease the friction points with Moebius 8200 or 8201. Leave in position 1.
2. Replace the electronic module, making sure the two contact springs are properly located. Grease friction points, but not electrical contacts.
3. Replace the combined stop lever and switch, greasing the point of contact with the post on the setting lever. Replace the positive battery bridle and the cannon pinion. Grease the friction drive, and lubricate between the cannon pinion and the hour wheel.
4. Replace the bottom plate after inspecting the jewel holes.
5. Grease and replace the stem, checking its three positions. Leave it in position 1.
6. Replace the five stators. (Four are identical, the fifth, which is the 60-second counter, has a slightly different shape.) The coils

could be replaced now, or after step 21 – certainly the latter would render the coil less vulnerable.

7. Lubricate the bottom pivot holes of the bottom plate with Moebius 9034.
8. Replace the trains for the watch, the 60-second counter and the 60-minute counter.
9. Replace the train bridge. I find a clean, fine oiler or pricker about the only tool for manipulating pivots into their holes. With all pivots safely located, tighten the two bridge screws.
10. Oil the upper holes for the train with 9034.
11. Replace the 60-second chronograph wheel, after lubricating its front pivot with 9034.
12. Replace the upper train bridge, and secure with its two screws.
13. Oil the two pivots with 9034.
14. Replace the circuit distance piece.
15. Replace the three green coils, making sure they are correctly oriented, and secure with one screw each.
16. Oil the bottom pivot holes for the 12-hour train. Oil the front pivot of the 12-hour counting wheel, and replace the stator, rotor and 12-hour recording train.

17. Replace the 12-hour train bridge, and oil the top pivots with 9034.
18. Replace the red 12-hour coil, which is to the top right of the watch with the button pointing to the left.
19. Replace the stator for the $\frac{1}{10}$-second train, and the train after oiling the front pivot of the $\frac{1}{10}$-second counting wheel.
20. Oil the top pivots with 9034.
21. Replace the $\frac{1}{10}$-second red coil. If not already done, replace all five coils.
22. Check that the coils are facing the right way.
23. Replace the additional printed circuit board, making sure it locates properly over the coils, and secure with its six long screws. One secures the circuit distance piece. Take great care not to damage the coils.
24. Replace the upper magnetic screen and secure with its two screws.
25. Replace the battery, negative facing down.
26. Replace the two battery clamp screws, but leave them sufficiently loose to insert the clamp.
27. Insert the battery clamp and tighten the screws.
28. Grease the teeth of the date indicator, then replace it, locating the date jumper.
29. Replace the three small date-indicator maintaining plates.
30. Replace the dial.
31. Turn the button so that the date just starts to turn, then replace the hour hand at 11.30.
32. Turn the hour hand to any convenient hour, and replace the minute hand.
33. Replace the small continuous seconds hand, and check that all three hands clear each other and the dial.
34. Temporarily replace the seconds-recording hand (the smaller of the two sweep hands) just to see what is happening.

35. With the stem in position 1, press the 'A' pusher with blunted pegwood to start the chronograph. If it doesn't start, push it again. Once started, press it again and it should stop.
36. Remove the seconds-recording hand, and replace the 12-hour counter, the 60-minute counter, the seconds counter and $\frac{1}{10}$-second counter to any convenient register – it need not be zero.
37. Now press the 'B' pusher, and the hands will rotate and stop randomly. To correctly zero the hands, proceed as follows:
38. Put the stem into position 2: repeated pushes of the 'A' pusher then sets the hour counter to zero; a quicker adjustment is made by holding the pusher down. Successive pushes of the 'B' pusher will zero the 60-minute recording counter.
39. Put the stem into position 3, and successive pushes of pusher 'A' will zero the seconds counter. Successive pushes of 'B' will zero the $\frac{1}{10}$-second counter.

The hands are now zeroed, and the chronograph will function normally, with start and stop on pusher 'A' and with the return-to-zero on pusher 'B'.

Electrical Tests for the ETA 251.262
Consumption with the stem in position 1, chrono engaged: 8.2μA.
Consumption with the stem in position 1, chrono stopped: 5.8μA.
Consumption with the stem in position 3, without motors: 3.7μA.
Lower working voltage limit: the watch should function to just below 1.3V.
Resistance of the red coils, 1.5 to 2.5kΩ.
Resistance of the green coils, 1.0 to 2.0kΩ.

If electrical tests are passed, case the movement.

14 THE SEIKO KINETIC

When battery watches first appeared, the question, 'What if your battery fails in the middle of nowhere and you can't get to a shop?' was often asked. Well, Seiko have come up with an alternative to a battery: a miniature electric generator in the watch which charges a capacitor. They called the watch the Seiko Kinetic, the word 'kinetic' meaning 'due to motion' – hence kinetic energy.

The principle of an eccentric weight as used in an automatic watch is utilized, but instead of winding a mainspring through reduction gears, there is a step up in gear ratio through the oscillating weight wheel; this drives the pinion of a generating rotor causing it to turn, generating electricity in a generating coil block. For one turn of the oscillating weight, the generating rotor turns very nearly ninety times. This charges a capacitor which, when fully charged, will drive the watch for in excess of seven days.

Apart from the obvious advantage of not having to change batteries, the watch is more environmentally friendly because the life of the capacitor is indefinite so there is less pollution. It is easy to service and has all the benefits of quartz watch timekeeping.

To start a stopped watch, swing the watch about a hundred times, as illustrated in Fig. 176.

The state of charge of the calibre 5M series is revealed by pressing a button at the 2 o'clock position and observing the motion of the seconds hand. A rapid seconds-hand movement will indicate the charge, after which the seconds hand stops for the appropriate number of seconds so that the correct time can still be shown. The table overleaf shows the action of the seconds hand for the duration of the charge,

and the number of swings of the watch, assuming that the capacitor is already charged to 0.5V.

In the calibre 5M42A, the button has three positions. Position 1 has no function, and is the position when the watch is normally worn. Position 2 allows a rapid date change, forwards only. Position 3 is normal hand-set and gradual date change; it is possible to advance the date without turning the hands through twenty-four hours. After the date has changed, simply turn the hands backwards by about three hours and forwards again. There is a stop-seconds with this calibre as well.

OVERHAULING THE 5M42A

Dismantling

1. Undo the bracelet and remove the case back. The oscillating weight and two coils can be seen. (Fig. 177)

Fig. 176 Swinging a Kinetic to charge the capacitor.

Number of swings	Duration of charge	Rapid movement of the seconds hand when the indicator button is pushed
100	Approx 6 hours	5 seconds
400	Approx 2 days	10 seconds
700	Approx 4 days	20 seconds
1100	Approx 7 days	30 seconds

2. To remove the stem, pull the button out to position 2, and push the dimple in the setting lever with the point of the tweezers. It lies under the contact for checking capacitor charge condition. (Many Seiko calibres are removed from the case in this way.)

3. Lift the movement out of the case.

4. Replace the stem. A positive push on the stem is enough to locate with the setting lever.

5. Bring the hands on top of one another, cover them with tissue paper, and lift them off. Leave the stem in position 3 for this.

6. Remove the dial by lifting carefully; there are no dial screws.

7. Remove the dial washer.

8. Remove the holding ring for the dial.

9. Holding the movement in your left hand, slacken the screws holding the date dial guard.

10. Remove the day-corrector setting wheel and the intermediate wheel for calendar correction.

11. Remove the date dial.

12. Remove the date jumper, the hour wheel and the date driving wheel.

13. Turn the watch over and remove the screw holding the oscillating weight, then remove the weight and the oscillating weight wheel. For this step the movement could be held in a movement holder.

14. Remove the two capacitor clamp screws, the clamp and insulation. Take care when lifting the capacitor insulation that the holes for location don't tear.

15. Remove the capacitor unit by prising against the main body. Do not lift by the capacitor lead terminal.

16. Remove the two circuit-block cover screws and the circuit-block cover. Take care with the screw close to the coil. The screwdriver blade could easily damage the coil.

17. Carefully remove the circuit block. It can be tight on its location posts.

18. Remove the two oscillating weight bridge screws and the bridge.

19. Remove the generating coil block.

20. Remove the intermediate wheel for the generating rotor.

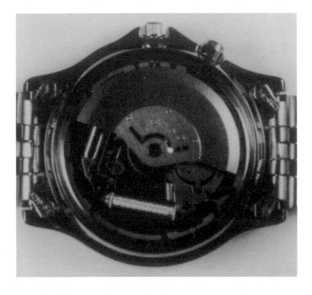

Fig. 177 Inside the back of the Kinetic 5M42A.

21. While holding down the generating stator with pegwood, remove the generating rotor.
22. Remove the generating rotor's stator.
23. Remove the coil-block screw, and the coil block.
24. Remove the single train-wheel bridge screw.
25. Remove the train-wheel bridge.

Study the layout of the train and hand-setting mechanism for ease of reassembly. (Fig. 178)

26. Remove the train wheel setting lever.
27. Remove the capacitor connection.
28. Remove the train. There is a brass fourth wheel which carries the centre-seconds hand, a third wheel with three holes (pinion down), a fifth wheel pinion up, a rotor, a minute wheel with coarse teeth, and a setting wheel with a pipe sticking up.

Fig. 178 Layout of the train.

29. Remove the switch lever.
30. Remove the yoke.
31. Remove the setting lever.
32. Remove the rotor stator.
33. Remove the stem and sliding pinion.
34. Remove the centre wheel and pinion.

You are now left with the empty plate. Parts which must not be cleaned in the cleaning machine include: rotor, coils, circuit board, parts made of synthetic material (plastic), and the capacitor unit. Clean the capacitor unit with Rodico, and the remainder of the above with isopropyl. The rest of the watch can be cleaned in normal cleaning fluids.

Check the train coil block resistance, which should be between 1.7 and 2.1kΩ. The generating coil block resistance should be between 280 and 380Ω.

Reassembly
1. Lubricate the clutch arrangement in the centre wheel and pinion with 9010 (cannon pinion), and replace.
2. Lubricate with 9010, and replace the stem and sliding pinion. Leave the stem in position 1.
3. Replace the rotor stator for the step motor.
4. Replace the setting lever, and lubricate with Seiko watch oil S-6 (or pressure-resistant oil or grease).
5. Replace the yoke and lubricate with S-6.
6. Replace the switch lever, lubricating with S-6.
7. Replace the train, starting with the setting wheel, the long pipe uppermost and the pinion towards the plate; then the minute wheel with the small pinion down; next the rotor, holding the stator in position with pegwood while doing this; the fifth wheel, pinion up and engaging with the rotor pinion; the third wheel with its three holes, pinion down; and finally the brass fourth wheel (centre-seconds wheel).
8. Replace the capacitor connection.

9. Replace the train-wheel setting lever over its post locating one end with the yoke.
10. Replace the train-wheel bridge and screw. Make sure that all the wheels are properly located and have endshake.
11. Lubricate the pivot holes with Moebius 9010.

    As the train bridge is held by one screw only, do not operate the stem and hand-setting yet.
12. Replace the step-motor coil block, and one coil block screw at the opposite end to the circuit connection. The coils are not interchangeable, and the circuit connections face up.
13. Replace the generating stator.
14. Holding the stator stationary with pegwood, replace the generating rotor with the pinion facing up.
15. Lubricate, with 9010, the three legs on the clutch spring of the intermediate wheel for the generating motor, the tips of the teeth and the centre of the wheel, and replace in the watch with the pinion facing up.
16. Replace the generating coil block.
17. Replace the oscillating weight bridge. Secure with the two screws nearest to the centre of the watch. The outside screw hole is for the capacitor clamp.
18. Lubricate with 9010 the two pivots in the oscillating weight bridge and the ball bearings.
19. Inspect and replace the circuit block containing the IC and crystal. Take care locating the circuit block over the location posts.
20. Replace the circuit-block cover and two screws. The third screw is for the capacitor clamp. Take care not to damage either coil with the screwdriver.
21. Try the hand-setting mechanism:
    Position 1: the button must be free to turn.
    Position 2: the clutch wheel will move out slightly, but still the button must be free to turn in each direction.
    Position 3: the clutch wheel must move 'in', a slight resistance should be felt as the

button is turned, and the minute pinion should be seen to turn.
22. Return the button to position 1.

Measuring the Total Current Consumption
   (a) Temporarily replace the capacitor clamp screw 'B' (Fig. 179).
   (b) With an external supply of 1.55V and a gate time of 10 seconds, feed the capacitor input terminals, observing that the rotor is turning at one-second intervals. This can be seen at 'A', through the jewel hole.
   (c) Wait 30 to 40 seconds for a stable measurement, then note the consumption: it should be less than 0.7µA.
   (d) Remove the capacitor screw temporarily replaced for the check just made.

23. Replace the capacitor unit, making sure the capacitor lead terminal is correctly located.
24. Replace the capacitor insulation, locating it over its two steady pins.
25. Replace the capacitor clamp and its two screws.
26. Lubricate the tips of the teeth of the oscillating weight wheel, and replace. The wheel is dished and is replaced as an upside-down saucer.

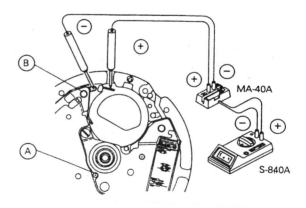

Fig. 179 Testing the total consumption of the Seiko Kinetic using the Seiko S-840A and MA-40.

27. Locate the oscillating weight and secure with its screw.
28. Turn the watch over and lubricate the bottom pivots.
29. Replace the hour wheel, the date driving wheel, the date jumper intermediate wheel for calendar correction, and the day corrector setting wheel. Part of the calendar work is equipped for day/date, although on this calibre they have no function.
30. Replace the date dial, and load the date jumper spring.
31. Replace the date dial guard and its three screws; tighten these lightly at first. If the date dial is free, tighten them fully.
32. Test the mechanism with the stem in positions 2 and 3. In position 3 the date should move forward one date every twenty-four hours. Return the stem to position 1 again.
33. Supporting on the oscillating weight screw, replace the holding ring for the dial, the dial washer and the dial.
34. Put the stem into position 3 – that is, two clicks out.
35. Turn the button clockwise until the date is seen to begin to advance. Replace the hour hands to indicate 11.30. The hour hand should be pushed on so that it is flush with the top of the hour wheel and parallel with the dial.
36. Turn the hand to a convenient hour so that it is pointing exactly to the hour.
37. Replace the minute hand so that it is flush with the pipe over which it fits – and in any case, the hour wheel must be left with endshake (though the amount isn't critical).
38. Replace the seconds hand so that it rests on the minute mark at the top of the dial. (Hopefully, with a well printed and centred dial, the seconds hand will rest on a minute marking as it advances around the dial.)
39. Remove the stem and case the movement, making sure the contact between the movement and the case is not damaged.
40. Replace the stem.

41. Temporarily replace the case-back after lubricating the thread and water-resistant seal with silicone grease.
42. Swing the watch to establish that it is functioning, then wear it for sufficient time to fully charge the capacitor and to complete electrical tests. In my workshop I have a Bergeon 5802 which is made for positional testing, and to wind automatic watches. It is ideal for winding the Seiko Kinetic without wearing it.

Seiko say that the monthly rate at normal temperatures is less than 15 seconds a month, the operating voltage of the capacitor is 0.5V to 2.3V, and that consumption is measured with a gate time of 10 seconds and should be less than $0.7\mu A$ for the whole movement with the voltage supplied from the capacitor. The consumption for the circuit block alone is less then $0.4\mu A$.

Checking the Automatic Generator System
Set the test equipment to read voltage, and apply the test probes as shown. You should have a value of between 0.5 and 2.3V with a going watch. (Fig. 180)

43. Finally, tighten the case back.

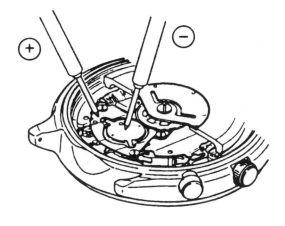

Fig. 180 Checking the automatic generating system.